THE ENERGY RIP-OFF

THE

50 YEARS 1924-1974

INTERNATIONAL PUBLISHERS
New York

ENERGY RIP-OFF

CAUSE & CURE

GUS HALL

First Edition
Printed in the United States of America 209

Library of Congress Cataloging in Publication Data

Hall, Gus
The energy rip-off.
Includes bibliographical references
1. Petroleum industry and trade—United States
2. Energy policy—United States. I. Title.
HD9566.H25 338.2'7'2820973 74-8897
ISBN 0-7178-0421-6

CONTENTS

CONTENTS

THE ENERGY RIP-OFF

INTRODUCTION

Your electric bills are monstrous; gasoline is seventy cents a gallon; and you're told to keep your house colder in the winter and hotter in the summer to "save energy." Who's to blame for these disasters? We can blame earthquakes on pressures in the earth's crust; floods on excessive rains and on the government for not building dams; hurricanes and tornados on air currents; flu on a virus; a broken leg on an accident. But what brought on the energy crisis? Some say shortages. But what brings on shortages? *Who shut off the spigot?* The answer to this very important question affects the lives of all of us.

The Watergate hearings have exposed the systematic effort by Nixon and the people around him to cover up the corruption and scandals of his administration. These are serious matters, but they are small potatoes compared to the massive cover-up of the scandal, the crimes behind the energy crisis.

There is a high-level campaign to blame you, the public, for this crisis. *You* are the culprits. Some come right out and say so, some attack the flank. "You were warned of an oil shortage in the thirties. It never came. After the Second World War the alarm sounded again, yet we doubled our energy consumption. Now you are being warned again of an energy crisis. You must be wondering if it is for real. It is. This time the wolf is really here. *He is us*—the government, industry, the *American* people."* That is how the liberal senator, William Proximire, pointed the finger at the people of Wisconsin. It is impossible to blame anyone specifically for "we are all to blame."

In its editorials, the *New York Times* hammers at the same theme: life for the vast majority (of Americans) is comfortable and profligate and overindulgent, eroded by easy credit and easy living. The word "profligate" tells you you are extravagant, a squanderer, and wasteful. In another editorial the *Times* editors take one more shot: the real crisis is failure of government and of the people the government represents. The obvious aim is to blame anyone as long as no one points the finger at the real culprits. In these editorial statements there is not one word about Exxon, Chase Manhattan Bank, Mobil, or Con Edison. This is a massive cover-up for the biggest rip-off in the history of our country.

You, whose electric bills have shot up 400 percent; you, who cannot make the ends of paychecks meet; you, the unemployed worker laid off because of the energy crisis; you, in unheated slums and barrios because of the racist system that put you there; you, the poor farmer being squeezed from all sides; you, the pensioner whose meager checks buy food for Mondays, Tuesdays, and Wednesdays only—*you* are the cause of the spiral of inflation and the energy crisis. You are too comfortable and profligate. You are eroded by easy credit and easy living. You are extravagant, squandering, and wasteful. One cannot help wondering what the two old couples who froze to death this winter, one in New York and one in Wisconsin, would have said about comfortable living, about being profligate. In both cases they died because the energy corporations coldbloodedly turned

* Newsletter: U.S. Sen. William Proxmire Reports to You from Washington, December, 1973.

off their source of heat, and then excused their action with the phrase, "We are in the business of making money." That goes to the very crux of the problem. Life in the United States is dominated by corporations who are in the business of making money. The business of making money is in a deepening contradiction with the human desire to make life more comfortable, to preserve human life.

Senator Proxmire states cheerfully: "You have a chance to see how people lived before with shortages."* Maybe that would be a unique experience for U.S. senators. For most workers, farmers, Blacks, Chicanos, Puerto Ricans, Indians, and Asian Americans—for most of the students and youth—it is not a unique experience. Living with shortages is a familiar part of their lives.

In this book we will examine the energy crisis, its causes and effects. We are going to examine it as symptomatic of serious and deep-seated problems, to quote another *Times* editorial. The examination must start from the premise that the energy crisis and the rising spiral of prices, rents, and taxes are not momentary waves that will subside. They are not passing nightmares. There will be periodic ups and downs, but the longer-range outlook is more of the same. There will be ebbs and flows, but on the road ahead there will be crises and more crises. In one form or another continued energy crises, food shortages, price rip-offs, criminal pass-on-the-cost additions to electric bills, rents, and taxes are going to be a feature of life in the United States from now on. They are symptoms of a new stage in the decay of capitalism. It is a form of corporate hijacking.

* Ibid.

part 1

• THE PROBLEM
• THE ISSUES
• THE ASSESSMENT

1 • "We Would Like You to Know"

The Exxon TV advertisement that closes with, "We would like you to know," is very instructive. In a sense they are saying: "We will tell you only that which we think you, the public, ought to know." What the corporations would like you to know about their operations is—nothing. What they do *not* want you to know is the most important story of this moment. It is a story of an unprecedented, colossal, criminal, corporate rip-off. It is a story of a crime and its cover-up.

Speaking off the record an executive of one of the largest oil corporations confessed: "Look, all oil companies plan 10 years in advance as to what the marketing resources and production capabilities will be. That is standard procedure. With or without the cutoffs of Arab oil they knew what was going to happen."

Ten years ago the computer tapes in the offices of Exxon, Chase Manhattan, Mobil, Gulf, Con Edison, the Standard Oils of California, Ohio, New Jersey, and

Indiana might have typed out the words: "Projection—1973 energy crisis—shortages and price increases—can result in a fivefold increase in oil corporate profits." But don't start looking for the tapes because they, like some others, most likely have been erased or destroyed. The tapes projected a plan of action that would create shortages.

The computer projections of shortages had nothing to do with a possible depletion of natural resources of oil. They had nothing to do with the coming oil embargo by the Middle East oil producing countries. The computer tapes recorded and projected a corporate, criminal conspiracy in the interest of maximizing profits. The plan was rather simple: create shortages and use the shortages to hike prices.

The 1973-74 energy crisis is witness to the deadly accuracy of the projections. There are "shortages," an "energy crisis,"and huge, windfall corporate profits.

What was the nature of the data that would have been fed into the computers? The basic data were the projected policies the oil corporations were to follow during the coming years. These policies included refusing to build any new oil refineries in the United States; phasing out older and smaller refineries by dismantling or simply closing them down; gradually cutting off the supply of fuel to the smaller, independent refineries, forcing them to either sell out or close their doors; refusing to drill for new oil wells on U.S. soil; capping older oil wells and drilling new ones next to them because the government depletion allowance was higher on the new wells.

The data also would include existing and projected governmental policies: bigger depletion allowances, bigger tax loopholes for the corporations, and even bigger ones for their overseas operations; continuing an embargo on overseas oil, permitting only a very limited amount to come in. These policies turned the State Department into a vehicle for corporate overseas operations, firmly committed to keeping a massive military presence in all corners of the world, not only as a threat, but to be used whenever necessary and advisable.

The above made up one side of the equation fed into the computers. The other side of the equation includes the following: a projection of the growth in the use of automobiles, including the increase in their gas consumption; increase in the use of jet and diesel fuel; in the use of oil, gas, and electricity in homes and apartment buildings; and in the use of oil and gas by industry.

Based on the corporate policies of expansion and mechanization, which were fed into the computers, they correctly calculated that the refining capacity in the United States would increase by about 3.7 million barrels per day, while the demand for oil products, including gas and oil, would be increasing at the rate of 6.3 million barrels per day.

The mathematics of such an equation do not require a computer. Supply and demand were on a collision course. Thus, while the energy crisis was a surprise to most Americans, it was no surprise to corporation and bank executives. They were ready for it. That is the story Exxon would *not* like you to know.

The Monopolies' New Technique

Shortages are a new tool of corporate robbery.

When the line between supply and demand is kept thin, shortages, whether created or real, become a weapon for maximizing profits on all levels. Under such conditions even a threat of shortages is effective. Corporate officers initiate rumors about possible shortages and then use them to jack up prices. Shortages become an inflationary pressure at all levels of the production and circulation cycle.

Why corporations are now able to create and use shortages is a very important part of the crisis story.

The computer's-eye view of reality tends to be mechanical. Its calculations are correct, based on the data it is fed. In the pages that follow we are going to fill in where the computer tapes left off—the reality the computers did not get.

The data, factual and political, behind the energy crisis are like a showcase; an exhibit in which all of the problems, processes, and contradictions that are present in life generally are sharply focused. To study the energy crisis is to study capitalism in the United States.

In a January 13, 1974, editorial, the *New York Times* reflects on this fact: "As the lines at the gas stations get longer and prices everywhere go higher it becomes clear that this country is facing more than problems of energy and inflation. These are symptoms of something more serious and deep-seated." The crisis of energy and inflation is indeed a symptom of something more serious and deep-seated. The editorial worries that there is "an uneasy reckoning with the American way of life." When *Times* editorials refer to the "American way of life," it is obvious they do not have the evils of the capitalist system in mind. They are craftsmen at hiding such evils with euphemisms like the people, the nation, the public, a worldwide problem. Like all mouthpieces of big business they write only what they would like you to know, and that does not include the corporate criminal conspiracy responsible for the energy crisis.

Was there a conspiracy to create an energy crisis? Did the corporate executives get together in a secret meeting to plan for the energy crisis? As expected, the *Wall Street Journal* came to the defense of the oil corporations. Having no intention of dealing directly with the charge, the editors respond with belittling and backbiting. The editorial is entitled, "The Great Oil Conspiracy":

> At cocktail parties, on railroad platforms, and in elevators, we have encountered an astonishing number of people who know all there is to know about the energy industry and how it conspired to cook up the energy crisis. They tell us the oil tycoons acted on an opportunity they contrived to push up prices, get the Alaska pipeline approved, and squeeze those pesky little independents out of business.
>
> Conspiracy theories, of course, are inherently appealing in that they can be understood without regard to any messy little details.*

It is possible the people do not know the messy little details. But what is certain is that the editors of the *Wall Street Journal* are committed to covering up the messy big criminal conspiracy by the oil corporations. Actually, the *Wall Street Journal* states the case quite well. The shortage of gasoline, fuel oil, and natural gas has been spawned by the handful of monopoly corporations who have an iron grip on the production and distribution of energy. And the editors of the *Wall Street Journal* are right—the aim of the conspiracy was to contrive an atmosphere in which they could blitzkrieg a crash program to push up prices, and to rush congressional approval of the Alaska pipeline. Yes, and to get rid of the pesky little competitors and pesky environmentalists, to get billions from the government for research, to get new tax reductions, new special exemptions, etc.

The Teapot Dome scandal involved an oil corporation stealing oil from lands controlled by the United States Navy. The name of the corporation has changed, but the oil under United States Navy control is again being stolen today. The theft, like every crime

* *Wall Street Journal*, 15 January 1974.

of the conspiracy, is defended because there is "an energy crisis."

The *Wall Street Journal* tried to belittle the conspiracy, but the truth is that the energy crisis occurred. The oil corporations achieved most of the goals of their criminal conspiracy. The aim was windfall profits, in that sense the conspiracy was a great success.

The Conspiracy and the Criminals

The energy crisis is the product of a specific conspiracy—in the field of energy—operating within a conspiracy—an economic system that is itself a conspiracy against the people. The oil corporations and the *Wall Street Journal* do not like the word, but the truth is that it is not only a conspiracy, it is a massive criminal conspiracy.

Did the criminals meet secretly behind locked doors to draw up their plans? Most likely not. They did not have to and that is what makes this development even more serious. They worked on a daily basis in a normal business system of interlocking interests.

Some fifty years ago, as a result of the Teapot Dome scandals, a few men went to prison for short periods, and the scandal was soon forgotten. There was a trial there were hearings; but the economic processes, the political system that gave rise to the scandal, was never touched. In fact the system and the inherent corruption in the system were carefully covered up. The economic processes that gave rise to the Teapot Dome scandals are present and alive in today's energy crisis.

The Exxon "we would like you to know" commercial inadvertently gave the clue to the process. In the TV film clip the diver shows how the steel foundations of the undersea oil pumps attract small plant life that in turn attract small fish who attract bigger fish, who attract even bigger fish. That is a very good description of the economic process of capitalism. There is, of course, one big difference. As a rule nature replenishes the smaller fish so there is a continuous cycle. Under capitalism the "pesky little competitors" are swallowed up and the biggest fish get bigger. Just as the smaller fish around the underwater steel pilings of Exxon's oil wells have been eaten, so the Exxons of the energy complex have devoured their competitors one after another, leaving a handful of greedy, crafty, brutal, rich families who have stolen the natural resources that by all standards belong to the people. They have cornered the markets and now sell the oil products back to the people who owned the oil resources in the first place. Oil corporations need no secret meetings; the whole monopoly is itself an intertwined, intermeshed conspiracy. Exxon, the largest industrial corporation in the world, has executive directors interlocking and overlapping with the boards of directors of Chase Manhattan, Morgan Guaranty, and the Chemical banks. In turn, these three banks have interlocking relationships with their people on the boards of directors of Cities Service, Mobil, Hess Oil, Continental Oil, Standard Oil of Indiana, Atlantic Richfield, and many more. Chase Manhattan Bank, the chief bank of the oil industry and the Rockefeller family, unites on its Board of Directors J. K. Jamieson and John H. Loudon, chairmen respectively of Exxon and Royal Dutch Shell, the biggest of the Seven Thieves International Petroleum Cartel, the two chief international plotters of the energy shortages. This is not an exceptional situation. These octopus-like energy conglomerates have in their iron grip the total energy complex. The U.S.-based energy octopus, with but seven corporate tentacles, almost captured the entire resource of energy in the nonsocialist world. It is the closest thing to a world monopoly in history.

The interlocking is not limited to boards of directors. With their people in the federal government they interlock in what are called joint ventures, to control the bidding for government-offered oil fields. They effectively lock out the smaller competitors and set their own price. Their men in the federal government deliberately make the lots so large that no small competitor can afford to make a bid. Using bids for offshore oil lands the corporations are continuing to steal the natural resources that belong to the people.

Reporting on a study made by the national Marine Engineers Beneficial Association, Jesse M. Calhoon stated to a Senate subcommittee that the joint ventures provide opportunities for "exchanges of information, division of marketing priorities, perhaps some production planning and perhaps a general form in which a climate of unanimity with respect to such problems as *scarcity, prices, political associations and other pertinent affairs* can be developed."* (Emphasis mine—GH) That is a diplomatic way of saying that the system of enmeshed relationships provides an opportunity for a continuing conspiracy against the people.

By using the crisis to drive most of the smaller, pesky independents out of the business, the big corporations were able to move to the next step. They are now dividing up the market into exclusive territories. Gulf Oil announced that it is closing all of its gasoline outlets in the state of Georgia—Hess in turn closes its stations in another state, leaving that state for Gulf Oil. It is not just Gulf and Hess. All of the major oil corporations are now involved in dividing the states for exclusive exploitation and to eliminate competition wherever possible. When the territory is divided

* Hearings before the Special Subcommittee on Integrated Oil Operations, U.S. Senate Committee of the Interior and Insular Affairs, 12 December 1973.

the nature of the competition changes. The corporation with a monopoly on the outlets then cuts down on the number of gas stations, cuts down on services, and thus is in a position to regulate and determine prices for gasoline and other services. Whatever one calls it, and no matter what phase of the operation you study, it is a conspiracy against the people.

Oil as it comes from the wells is raw material for gasoline, fuel oil, and other oil products. It is the refineries that turn oil into usable energy. The refineries are mainly owned, controlled, and run by the interlocked, exclusive energy conglomerates. The gasoline shortage began to appear in 1971. Since then the top five oil corporations cut off oil to the last of the independent refineries, forcing some to close their doors during the height of the crisis. And though the corporate executives knew ten years ago that shortages would appear, at this writing there are no new refineries under construction in the United States. Some old refineries are being rebuilt, but the corporate ban on new construction remains in force. The energy crisis is working so well that Exxon and Shell even canceled plans they had on the drawing boards. (It takes from three to four years to build a new refinery.)

In order to bamboozle the public, some oil corporation executive makes an announcement every so often about plans to build a new refinery. Donald O'Hara, vice president of the National Petroleum Refining Association, said: "We call these things press-release refineries . . . they are worthless until a company actually breaks ground."* But the corporations are not breaking ground. Exxon, the biggest thief, has for years released press statements about building a large, new refinery in Bayway, New Jersey. When pressed for specifics a spokesman declared: "It was still being

* *Wall Street Journal*, 14 November 1973.

designed and the actual decision as to where to build is still being reevaluated." Shell Oil also has backed down on one of its press-release refineries, saying: "But even if we were to obtain a permit to go ahead we would not start construction until we have a better picture of what lies ahead. . . . There is to be no further expansion until we know where we stand." That is the voice of leading members of the conspiracy, right in the midst of the energy crisis. It is the biggest rip-off in the history of the United States. The airline hijackers grabbed airplanes, the oil conglomerates are hijacking a whole people, a nation.

The Long Arm of Conspiracy

How does it happen that there is a shortage of natural gas exactly when there is a gasoline shortage? For the very simple reason that the gas wells and gas distribution are owned and controlled by the same few corporations that own and control gasoline and oil. As they have not built refineries, they have also not drilled for gas. The demand was bound to surpass supply. That was the purpose of the conspiracy.

This fact begins to reveal the enormity of the energy rip-off. The same few corporations have in their grip the total energy complex, controlling most of the oil, coal, gas, and uranium resources. In all these fields they control most of the extracting and refining complexes, as well as most of the wholesale and retail outlets.

They have a controlling influence in both the Democratic and Republican parties. They have a controlling influence in such splinter movements as the New York Conservative Party and the Wallace movement in Alabama. They have a controlling influence in the federal government and in most of the state and city governments.

The Watergate hearings have provided a sidewalk superintendent's peephole into the overlapping and interlocking relationships between the monopoly corporations, the Republican and Democratic parties, and the federal government. The millions of dollars fed into Nixon's reelection fund by the largest corporations is but an indication of this influence. The Watergate hearings also exposed the process of monopolization in the structure of government. It parallels and is related to the process of monopolization in the corporate structure. The big corporations have monopolized the economic field, and the executive branch of the government has monopolized governmental activities, and between the two there is an interlocking, overlapping relationship. Thus, the control and direction of the country's affairs, economic and political, have increasingly passed to a state monopoly hub, greatly increasing the danger of a police-state setup. The oil corporations are in the very center of this hub, both as to the federal government and the corporate monopoly structure. On the list of the "Top 500" largest industrial corporations, Exxon is number one in assets and profits. Of the very top sixteen corporations in assets, eight are oil corporations. So when people talk about the need for a national energy policy, or when people say the government is responsible for the energy crisis, they are not referring to a neutral body. They are talking about policies and government bodies influenced, and to a large extent controlled, by these large corporations.

To change the energy policy of the federal government requires breaking up the interlocking grip the monopolies have on the government, which means electing an antimonopoly Congress. To elect an antimonopoly Congress requires building an independent political movement that is not influenced by the corporations.

The energy crisis forced everyone to take a deeper

look at our society—how it functions, its values and priorities—to ask what motivates and propels an economic system that creates such crises. What is the role of the government in all this? What is meant by a change in life-styles? What is the solution to the energy crisis, the food crisis, fertilizer crisis, and all the other crises? We will discuss these and other matters. But first let us look into the showcase, the oil industry. The oil corporations are, in a sense, a microcopy of the capitalist system. What you will see in oil, you will see in steel, coal, rubber, metal, machinery. What you see in oil you will see in the capitalist system. The same class forces, with the same class aims, operating in oil are present in every industry, in every phase of American life.

2 • Aim of the "Shortage"

The first thing to understand about the fuel shortage is that the oil companies wanted it.

For years their main economic objective has been to obtain higher prices for oil and natural gas. Higher prices for gasoline, the most profitable product, were hampered by the appearance and survival of market competitors, including small refiners and wholesalers, and European monopolies entering the U.S. market. So long as world supplies of fuel oil were ample the competing elements could stay afloat and restrict the ability of the larger corporations to raise prices. With the surge of inflation in the early 1970s, governments turned more to price and wage regulation. In European countries especially the oil monopolies were hampered by government price restrictions. And after Nixon introduced wage and price controls in

1971, possibilities of increasing prices of petroleum products in the United States were also limited. The oil corporations have been making huge profits. But the price increases and profits were somewhat limited. What the oil monopolies were after were profits and price increases without limits, without restrictions.

The price of natural gas meanwhile was regulated as a public utility by the Federal Power Commission, since most of it is distributed ultimately by regulated gas and electricity companies. The campaign for higher prices had to include political and economic pressure.

Nixon Toes the Line For the Oil Barons

The oil barons were high on the list of contributors to Nixon's campaign funds, to special funds for the "plumbers" and other dictatorial operations. According to a tabulation of Rep. Les Aspin (D.-Wisc.), 413 directors, senior officials, and major stockholders of 178 oil companies contributed, openly, close to $5 million to Nixon's 1972 reelection campaign. That was almost 10% of all the identified funds and did not include the large contributions of the Rockefeller clan and other well-known Standard Oil families not officially listed as major stockholders.

According to Aspin these contributions made it clear why "the administration attitude has been so consistently anticonsumer and pro-big oil. The big oil companies have Mr. Nixon in a double hammerlock. After paying for nearly 10% of the President's campaign it comes as no surprise that the oil companies are calling the shots."*

* *New York Times*, 2 January 1974

Nixon appointed 100%-promonopoly individuals to the Federal Power Commission, which made increasingly liberal grants of price increases for natural gas. And they are still calling the shots.

Nixon's appearance in Chicago on March 16 of this year was par for the course. The meeting at the Executive Club was arranged by the oil interests. The "spontaneous" question from the floor was, "Does this administration believe that the oil companies exploited a situation and pushed the prices to unbelievable heights and really took advantage of the American people?" Nixon's prepared answer was: "With regard to a *so-called* contrived energy crisis manipulated by the big oil companies . . . Now let me just make one point very clear . . . It would be very easy to just stand here and blame it all on the oil companies . . . But this is not a contrived matter . . . The energy problem has been one that has led to our economic difficulties." The only thing Nixon makes very clear is that he defends the oil corporations, even when there is no defense.

If it is not a "contrived matter" what is it? If not the oil trusts, who is responsible? In the same speech Nixon gives the oil corporations' answer: "And rather than to blame the big oil companies and say they could do something about it—they could if they had the oil . . . We should deregulate natural gas."—No limits on price increases—"We should move on the environmental field to relax some of those inhibitions." These are the only restrictions left.

To say the energy problem has led to economic difficulties is a statement in keeping with another President, Calvin Coolidge, who said, "The reason we have unemployment is because people are out of work."

There is no reasonable defense for the oil corporations. So Nixon keeps repeating that the big oil corporations are not to blame.

Starting in 1971, the oil companies took the lead in

propaganda about a world energy shortage. They claimed that in order to cope with the energy shortage vast capital investments were required, and the only way to obtain such capital was through higher prices and profits. A secondary theme was the demand for elimination of all environmental restrictions on the industry.

Thus, Maurice F. Granville, chairman of the board and chief executive officer of Texaco, urged the government "to consider the following measures to help solve the energy shortage: corrective measures, such as price deregulations, to permit free market forces to establish competitive prices for natural gas and petroleum products. Modification of sulphur and automotive emissions limits, which cause higher consumption of energy supplies.*

Granville added other proposals, but these two headed the list.

Standard Oil of California's officers in their 1972 annual report wrote of the "need to revitalize exploration in the U.S.," which, they claimed, required "positive steps in at least four ways. Energy prices must be permitted to rise to levels which fairly reflect increasing exploration and production costs. Tax policies must reflect the need to encourage the high-risk U.S. exploration effort. Leasing of federal lands, both onshore and offshore, must be accelerated. Finally, the nation must be certain that the steps it takes to preserve the environment are realistic and necessary."

To this direct propaganda for higher profits was added psychological warfare against the consumer by spreading fear of an energy shortage. To make the propaganda plausible dramatic shortages of heating oil were arranged in the winter of 1972-73 and given

* Texaco, *Third Quarter Report*, 1973

great publicity. The national media, for instance, carried pictures of a school in Colorado shut down for lack of fuel oil. Spot reports of small plants unable to get gas were given maximum coverage.

Since, in fact, stocks and production of fuel oil were at roughly normal levels the spot shortages clearly were deliberately organized by the petroleum monopolies, or were the kinds of incidents that occur frequently under conditions of capitalist anarchy. The big publicity and the ability of the oil cartel to get supplies where and when it really wants to do so, make the former explanation more likely.

Most powerful of all were the oil companies' efforts to create a changed supply and demand balance by curtailing supply relative to demand within the country. This was accomplished mainly by restricting the expansion of productive capacity, especially in refining, in a period of rapidly rising demand for petroleum products. Expenditures for plant and equipment of petroleum companies in the United States, which amounted to $5.08 billion in 1967, have declined continuously since then, as expressed in constant dollars adjusted for increases in cost of plant and equipment. By the second quarter of 1973 oil company expenditures for plant and equipment were at an annual rate of only $3.92 billion in terms of 1967 dollars, a decline of 23% in real terms.* Only one-tenth of this was spent on refineries. A considerably larger amount was spent on "marketing"—building excess gas stations to squeeze out independents, sponsoring radio and tv programs, paying for self-serving political ads and company-owned politicians' election campaigns.

The consumption of petroleum products increased 38% during this same period, and was increasing at

* *Survey of Current Business*, July 1970, September 1973: *Statistical Abstract of the U.S.*, 1972 (Washington, D.C.: Bureau of Census, 1972), No. 756, p. 474.

an unusually fast rate in the early 1970s. Therefore, there should have been a substantial increase in real capital expenditures. Thus the decline in investments, relative to requirements, was in the range of 40 to 50% between 1967 and 1973.

In part, the restriction of investment within the United States represented a shift of emphasis to foreign sources of oil and to foreign refineries at which the U.S.-owned monopolies could get bigger profits. By concentrating production more and more in the Middle East, and refining in small Caribbean islands, where they pay colonial wages and no taxes, the monopolies could more readily manipulate the supply of oil to the major industrial countries, including the United States, and create shortages at will.

Their capital spending abroad expanded from $3.0 billion in 1967 to $6.7 billion in 1973 and a planned $7.7 billion in 1974. The increase of more than 120% between 1967 and 1973 amounted to an increase of 69% in real terms. In 1974 foreign plant and equipment expenditures of U.S. oil companies are scheduled to rise another 21%, to $8.1 billion, considerably more than anticipated domestic expenditures. At the same time President Nixon demagogically talks about "self-sufficiency" in energy.

During the 1973 crisis, oil shortage propaganda became more and more intense despite the fact that the domestic supply-demand balance was actually improving, and new supply was increasing more rapidly than consumption. The oil monopolies were increasing production in the Middle East at a spectacular rate. Between July 1972 and July 1973 oil output in Saudi Arabia increased by an amount equal to the monthly production of Venezuela, one of the great oil producing countries. Production in Saudi Arabia approached U.S. and Soviet levels; and the projected oil company goals of more than doubling United States production levels by 1980, is scheduled to be matched and probably surpassed by Saudi Arabian production.

Meanwhile, the oil companies knew a showdown was coming in their negotiations with the Organization of Petroleum Exporting Countries (OPEC) over prices and terms of participation. They were aware that the pressures of the democratic and bourgeois nationalist forces in the Arab countries would soon put some limits to the reckless gutting of their natural resources. They also knew that the Pentagon policy of encouraging Israeli provocations in the Middle East created the danger of an Arab oil embargo, and that such an embargo would likely be more effective than the abortive efforts in 1967 and 1970.

Oil Shortage: Propaganda or Real

The actual imposition of a cut in production by the Arab countries in October 1973, together with an embargo on shipments to the United States and the Netherlands, created the conditions under which the oil companies could go into high gear with shortage propaganda.

Actually, there was never a shortage of crude oil in this country. The announced 25% cut in production in Saudi Arabia left output 8% above the average level of 1972; and some Arab countries did not participate in the cuts. And the oil companies furiously increased production in a number of non-Arab producing countries. Moreover, the oil cartel effectively overruled the Arab decisions as to which countries would be favored in getting the oil. By redirecting shipments from non-Arab sources, they saw to it that the first priority in oil supply went to the United States and the Netherlands, with cuts imposed primarily on France and Japan.

Meanwhile, the oil companies were actually cutting production here, while holding out for higher prices.

In the crazy quilt pattern of price regulation, oil from "new" wells was permitted to sell at any price, while ceilings were maintained on oil from "old" wells. Many companies simply capped old wells and drilled new ones right next to them. As a result crude production in the United States, at the height of the shortage scare, was 500,000 barrels per day below the 1972 level. Domestic refinery output was cut even more, by 800,000 barrels per day, during November and the first half of December.

And still the oil piled up. By mid-December, as lines were forming at gas stations, centralized petroleum stocks were 30 million barrels higher than a year earlier, not counting the huge hoards held by most of the large oil-consuming corporations, and widespread hoarding by capitalists able to afford big storage tanks.

The U.S. government and the oil monopolies then entered into a conspiracy to spread panic and aggravate the shortages. Vastly increased amounts of oil were commandeered by the Pentagon and shipped abroad to the U.S. fleets, military bases and puppet armies.

The Pentagon is the nation's largest single user of petroleum products. It now uses 637,000 barrels of oil a day. Twenty thousand barrels a day go to South Vietnam as Washington is still handling fuel procurement for South Vietnam's armed forces.*

The big oil monopolies have been going all out to supply oil to the Pentagon at the expense of the domestic supply. Thus *Businees Week* admits:

> Though it is heatedly denied by the oil companies, it is an open but unspoken secret that some of them are trying to help the military. Some product from Arab crude—originating in the Persian Gulf and North Africa—is being passed under the table to the U.S. military.**

* Jack Anderson, *New York Post*, 11 December 1973.

** *Business Week*, 1 December 1973, p. 19

And the federal government's Interior Department has been jumping on the Pentagon bandwagon. In early December 1973 it invoked the twenty-three-year-old Defense Production Act, passed during the Korean War but never before applied to oil. It ordered twenty-two oil companies to divert 825.9 million gallons of gasoline and other fuels from civilian markets and delivered it to military depots through January 16.

Children shivered through the winter in poorly heated schools, workers were laid off by the hundreds of thousands as a result of the shortage. Indeed, countless miseries yet untold are falling on the backs of the American people as a result of the contrived shortage, but the monopolies and the government make sure that the Pentagon gets its due.

The newly created energy bureaucracy, headed by a Wall Street investment banker and staffed by oil industry representatives, set up an allocation system aimed at creating a shortage for working people. In effect, industrial and most governmental users were allotted all the petroleum they wanted. But fuel distributors were ordered to cut their allotments to household customers by 15%, and to gas stations by 20%. In the absence of a rationing system, millions of people were forced into competitive scrambling for the reduced supply of gasoline. It was here that the effort was focused, as the refineries were instructed to cut output of gasoline and increase output of fuel oil. Nevertheless, the total output of gasoline remained close to the 1972 level.

The increase in stock was concentrated in fuel oil, so that the rationale for reducing gasoline output was completely destroyed. But the order remained in effect. Thus in the final analysis the cutoff of supplies to working people was accomplished by oil industry decision, enforced for the oil monopolies by the governmental "energy czar" William E. Simon. Such a complex operation could not have been carried out with guaranteed success by the monopolies themselves. The use of the government was needed to order compliance with the decisions of the most powerful monopolies.

If there were a real shortage, the logical thing would be to ration gasoline, and the public, in polls, showed they favored rationing and not price increases. The administration announced a "standby" rationing system, but Simon cynically declared that he would not put it into effect until people were forced to stand in line four hours to get gasoline!

The appointment of Simon was itself significant. The previous chief of the energy bureaucracy, former Colorado Gov. John A. Love, was not acceptable to the oil barons. Love, a minor lawyer and local politician by background, indicated preference for a rationing system that would make an attempt at fairness.

Simon, on the other hand, was a partner in the major Wall Street investment banking firm of Solomon Brothers, and boasted a personal share in the firm's profits of several millions per year. Solomon Brothers is the top firm floating tax-exempt municipal securities. Oil companies, among others, have been issuing tax-exempt "pollution control" bonds, a source of big extra profits to the bankers. Simon had the confidence of the oil barons and agreed with their policy goals, voicing their argument for freely raising prices as the way toward ample supplies.

Nixon transferred authority over fuel prices from the Cost of Living Council, whose members wanted to go slow in raising prices, to Simon, who proceeded quickly to give the monopolies what they wanted. Rep. Benjamin S. Rosenthal, (Dem., N.Y.) revealed that at least 52 key policy-making executives in the Energy Agency were oil company executives, temporarily transferring their profit-making activities from the company to the government payroll! Prices were increased at all levels—the oil well, the refinery, the wholesaler, and the retailer. Since the monopolies essentially controlled all levels, every increase represented an increase in their profits.

By April 1974 the average retail price of gasoline was up twenty cents per gallon, which amounted to an increase of more than 100% in the price received by the oil companies, before adding on the dealer's markup and the gasoline excise and local sales taxes. Further increases of ten cents or more a gallon were officially recommended by Herbert Stein, chairman of the President's Council of Economic Advisers. It must be stressed that the increases in retail prices went far beyond the higher cost to the monopolies of purchased crude oil. Some of the increases were so raw that they were publicly criticized by the Cost of Living Council director, John T. Dunlop, himself no slouch when it comes to raising prices for the monopolies.*

The American people, and the people of other capitalist countries, faced the prospect of a doubling of the cost of fuel within a brief period, and along with that a sharp rise in the difficulty of everyday living. In cities across the country workers were forced to line up for hours for small quantities of gasoline. The inconvenience and loss of working time was much more severe than during the real shortage, accompanied by a logical rationing system, during World War II.

How long this could be maintained was another question. By the end of 1973 tankers were lined up in New York harbor waiting for storage space for their cargo. The fiction of a severe shortage was being punctured, especially after the Arab countries announced their intention to increase production in 1974.

* *Journal of Commerce,* 3 January 1974.

"Shortage" Pays Off

But meanwhile the big rip-off had been put over, the operation, for the oil companies, was a success.

Independent distributors and at least 10% of all gasoline stations in the country were wiped out. Oil prices were way up and the ending of all price restrictions was well on its way.

Monopoly ability to restrict supply to the ultimate consumer was tighter than ever. The combination of the modern state and the monopolies proved more effective than the original John D. Rockefeller ever was in eliminating competition and robbing consumers.

The payoff in profits was colossal. The profits of the large oil companies in the fourth quarter of 1973, the first quarter of the supposed shortage, were up 75% from the fourth quarter of 1972, and three times as much as the quarterly rate of profits in 1963. That is, the ten year plan of the oil companies succeeded in tripling their profits. But that was just the beginning. Confident of their monopoly power, the Rockefellers and their allies set the most extravagant profit goals in history.

These were made public by the Rockefeller's Chase Manhattan Bank, leading bank of the oil industry. The bank estimated the profits, after taxes, interest, and depletion and other write-offs, of the thirty oil giants at nearly $10 billion in 1973. It set the goal of doubling these profits, to $20 billion in four years, by 1977; quadrupling them to $40 billion in 1981; and redoubling them again in a third four-year interval, to $80 billion in 1985. This represents an 18% compound rate of increase.

No industry, or even large company, has even come close to such a rate of profit over an extended period. The goal of the oil giants is to substantially exceed, by 1985, the record 1973 profits of all U.S. corporations.

Over the sixteen-year period, 1970-85, they aim for a total of $755 billion, or three-quarters of a trillion dollars. And this is in addition to hundreds of billions they plan to pay out as interest to bankers, and more hundreds of billions as booty to the huge corporate bureaucracy of the oil companies.

Altogether, the oil companies aim to reap in profits and other property income, in those sixteen years, an amount exceeding the U.S. gross national product for a whole year.

They anticipate such profits *if* recent peak rates in oil consumption continue without interruption; *and if* supply is always kept below this expanding demand so that prices can continue to be raised indefinitely.

Their argument for such out-of-sight profits is simple enough: they need them to finance the huge capital investments required to satisfy the oil requirements of this country and the rest of the capitalist world. But if the companies are granted such fantastic profits, they will *not* invest on the scale they project. Increasing portions of the profits will be used to swell still further the wealth of the billionaires and the hundred millionaires of the oil industry, to spread still farther the power the Rockefellers, Mellons, and other oil chieftains maintain over the entire economy and over the government.

Yes, big investments are needed to supply the world's energy needs. But the way to get them is not through private monopoly and private monopoly superprofits. The sources of energy are not inexhaustible. The easy-to-get energy sources are drying up. And though that is not the cause of the present shortages, it is a problem society will have to deal with from this point on.

When there were hundreds of companies extracting and refining oil it was difficult for them to agree on a single policy. Now there is but a handful. They are able to manipulate supply and demand, and even to use some inherent contradictions within capitalism against their competitors and against the people. They are able to create a shortage and turn it into a crisis which they translate into fantastic profits, and the crisis was clearly projected in the computer tapes of Exxon, Chase Manhattan Bank, Mobil, and Texaco over ten years ago.

3 • Shortages, Real and Man-Made

The world is not running out of sources of energy—tomorrow. Even hitherto traditional sources of energy are still far from exhausted. World production of coal and oil and natural gas output will continue to increase in the coming period. Within the United States there are still potential reserves of oil, gas, and coal.

The ability of a society to turn reserve energy into usable power is linked to the nature of its economic and political structure. A structure motivated by private profits does not measure up to the high-level energy consumption requirements of modern civilization.

For over fifty years the oil monopolies have periodically dragged out the bogey of vanishing reserves. Such campaigns have always preceded the drive for some tax concessions, for land grabs or price hikes. When they speak of only twelve years' "proven reserves" it sounds sinister. The proven reserves are those which the oil companies have not only already blocked out, but estimate that they can extract at existing prices so as to make the high profit rates they aim for. Professors Gramm and Davison of Texas A & M, recognized petroleum specialists, estimate that the actual reserves, "in these same reservoirs, in fairly well-defined locations and volumes," recoverable by known technology, *are ten times as large,*—120 years supply instead of 12 years. All that is needed, say the professors, is a higher price to give the oil monopolies "incentive."*

Consider this paradox. Reserves of gas are thought to be relatively larger than reserves of oil. Yet the shortage scare started in the winter of 1972-73 over natural gas. And the tight supply in fuel oil developed because utility companies, unable to get enough natural gas, switched to fuel oil, increasing their consumption at a fantastic rate.

What happened to the supply of natural gas? Let's see the explanation offered by Mobil, one of the big Rockefeller oil monopolies, in its advertisement of April 19, 1973, one of a series used to scare the public over the "energy shortage."

Headed, "The Unnatural Gas Shortage Revisited," the ad says:

> This nation has drifted into a critical shortage of natural gas, the cleanest-burning major fuel and one of the most convenient. Ironically the shortage is largely man-made. Made by the U.S. government, in fact.
>
> . . . All this trouble is downright unnatural when you consider that geologists and other explorationists believe there are substantial quantities of natural gas still to be discovered in the United States, especially offshore.

Why . . .? How did today's shortage come about?

Mobil's explanation: The government's regulation of natural gas prices has kept the price down. It is "impossible" for the companies to explore for new gas and increase production.

Now, let's look at the facts. Between 1960 and 1969

* *Wall Street Journal,* 26 December 1973

the Federal Power Commission held the price of natural gas steady. The gas companies made more and more money by increasing production. They increased output 66% in nine years. And their operating profit increased 77%.

And then Nixon came into office and proceeded to stack the Federal Power Commission with oil and gas company toadies who quickly handed out price increases. In four years, from 1969 to 1973, the average price per therm received from ultimate consumers increased 30%, and the price received at the field by the primary producers increased even faster. And it was then that the growth of production of natural gas slowed and ground to a halt.*

The Mobil theory of what was needed as an incentive to produce more is exactly the opposite of reality. Under Nixon the companies saw an opportunity to increase profits *faster* by producing less and increasing prices. And the more of that they got the more they wanted.

Yes, the shortage is man-made, but not just by the U.S. government. And not in the way Mobil claims. It is man-made by the men of Mobil, of Exxon, of Tennessee Gas Transmission, by the entire oil and gas monopoly, and the Rockefellers, Mellons, and Gettys who own it.

The possibility of cheap and abundant electricity which science and the new technology are now making a reality actually is an impossibility in the Exxon-GM world of today. They will make it as scarce and as expensive as possible if they have their way.

* Statistics compiled from *Statistical Abstract of the United States, 1973* (Washington D.C.: Bureau of the Census, 1973), p. 516, and *Survey of Current Business*, July 1973, November 1973.

Fuel for People or Fuel For Profits

The man-made shortage of oil and gas is a concoction of the owners of the monopolies to extort tens of billions in profits from the American people, and to make the price steeper each year.

The issue is whether to have unlimited, sky-high prices, and a declining supply relative to peoples' needs; or to have controlled, moderate prices, and production taken out of the hands of the monopolies and raised to whatever level the people need. Whichever way the issue goes it does seem likely that in the long run directly mined liquefied oil and natural gas will provide a declining share of the country's energy supply. But exploitation of other, and richer, energy sources has barely started.

The United States has enormous reserves of coal. But production of coal peaked in 1947, and even then it was only slightly above the level of 1918. Production of the high-quality Pennsylvania anthracite has virtually ended. Essentially, coal production has been held back by the competition of oil and gas which the monopolies obtained at very low cost through one-sided concession agreements in foreign countries, virtual freedom from taxes, and giveaway terms for access to U.S. resources. Furthermore, four of the ten largest holders of coal reserves are oil companies, and Exxon and Gulf, two of the biggest, are keeping their coal untapped, holding it back from production.

Of course there are environmental problems. Of course the destruction of large sections of Wyoming, West Virginia, and other states by strip mining must be stopped. Of course high-sulphur coal poisons the air. The solution, however, is not to stop mining, but to develop and apply the technology to cure these evils. The solution is to use the tens of billions of tons of low-sulphur coal being held back from production by Burlington Northern Railroad, Union Pacific, and other monopolies, and the vast government-owned reserves of low-sulphur coal also being held back from use.*

Despite the absence of an organized effort by the government, production of coal rapidly soared when the Arab oil embargo was put into effect. By mid-December 1973 coal output was running 15% above the previous year's level. Industry sources anticipated output at a record annual rate of 700 million tons of bituminous by the end of 1974, a gain of 110 million tons over the 1972 level. This would be the equivalent of 1.5 million barrels of oil per day, or 9% of U.S. consumption.

An energy report submitted by Dr. Dixy Lee Ray, chairman of the Atomic Energy Commission, recommended increasing coal production by 1980 to nearly a billion tons. But this was less than the recommended proposal of the Cornell Workshop group on coal which prepared the draft for Dr. Ray. This group recommended raising coal output to 1.2 billion tons in 1980, or double the recent level, and to 1.8 billion tons by 1985.

Companies Keep Lid on Coal Supply

Of course under conditions of monopoly control there are severe drawbacks to this actual and projected increase. Coal companies are extorting high prices for the extra coal. A propaganda campaign based on a "national emergency" was launched to deprive the coal miners of a decent settlement in the key 1974 negotiations. On the same phony basis, ecological and safety requirements are being disre-

* "Historical Statistics of the United States," *Survey of Current Business*, November 1973; "1971 Business Statistics," *United Mine Workers Journal*, July 1973.

garded, to enable coal to be burned without environmental safeguards. Dr. Ray's report also provides for speeding up the turnover of government land to the coal monopolies, and for spending $2.2 billion in federal research money for the benefit of these monopolies. To indicate the low priority of safety and labor benefit measures, the program considers it necessary to increase employment of strip mine workers by only 31%, and of underground miners by 50%, while production is to be doubled.* In everyday coal miner terms this means speedup, black lung, and mine accidents.

Considerable sums will be spent for development of new methods of gasification of coal. All this is discussed in the press as if coal was never gasified before, but for a century gas for cooking and lighting was produced mainly from coal. When and if natural gas supplies run short, or when natural gas prices reach a certain peak, this will certainly be revived. William P. Orr, president of Lummus Company, the engineering subsidiary of Combustion Engineering, stated that a commercial plant for the gasification of coal would be in operation by 1977 or 1978, years ahead of customary predictions. There is a problem of pollution in the manufacture of gas from coal, and we'll discuss a little later the prospects for solving this by underground gasification of coal.

Similarly, there are large potential reserves of natural gas in the United States. These range from two to six times current natural gas proven reserves. But the concern of the monopolies is profits, not the supply of energy for domestic needs for which there are potential reserves.

An independent study has observed that total proven reserves are only a small fraction of all the petroleum believed to exist in the world, or that might ultimately be recovered under different technical and

* *New York Times*, 6 January 1974.

economic conditions. Besides the oil in liquid form, the world has vast resources of petroleum in shales and tar sands. In fact an estimated 1.8 trillion barrels of oil, nearly three times the world's current proven reserves, are hidden in the shale rocks of the Rocky Mountains.

The problem has been separating the oil from the rock. Prior proposals required extensive strip mining that would devastate the countryside. Occidental Petroleum Corp. has a pilot plant for extraction of oil from shale in Colorado getting twenty-five to thirty barrels of oil per day. The special feature of this process is that all of the oil is extracted underground. The piles of shale dust which would otherwise create an almost insuperable environmental obstacle are avoided. Dr. Donald E. Garrett, research and development director of the company, claims that commercialization of this process should be possible in three or four years. Significantly, neither of the companies engaged in these efforts is part of the main monopoly groups in the oil industry.* The oil monopolies are deliberately exaggerating the difficulties associated with these processes. They are anxious to delay them as long as possible, in order to derive the maximum profit from price gouging and artificial shortages.

Nuclear Energy, the Answer?

Nuclear energy right now is coming into use on a larger scale. While attention has been focused—and understandably so—on the environmental issues, the power companies here and abroad have proceeded to order literally hundreds of nuclear power plants, and they are being installed at a very fast rate. In 1971

* *New York Times*, 16 December 1973; *Journal of Commerce*, 14 September 1973.

nuclear plants produced 2.4% of the country's electric power. In 1973 it was up to 5.2%. By 1985 it is expected to account for 45.2% of all electricity. This is not just something for the future. It is happening right now. During 1973 twelve nuclear power plants with a capacity of 10 million kilowatts became operable. During 1974 twenty plants with a capacity of 15 million kilowatts will become operable. These thirty-two plants can supply the equivalent of 800,000 barrels of oil per day, *more than half the increase in electricity consumption in 1974.**

True, there are serious technical problems with nuclear power plants which so far are keeping their operations down to as little as 50% of rated capacity.

Like almost everything else, the development of nuclear energy in the United States is hampered by monopoly profiteering at all levels. It is marred by profit-greedy companies which economize on safety precautions, do not care how many millions of fish they kill with hot effluent, do not worry about concealed long-term dangers in radioactive waste products, minimize sinister breakdowns in operating plants and threaten radioactive disaster.

Public vigilance and, even more, a mass struggle is necessary to protect the population generally and plant workers in particular from these dangers.

In the socialist countries the engineers and scientists spent much longer in experimental development of nuclear power plants than the profit-hungry corporations permitted their counterparts to do in the United States. Therefore, when large-scale nuclear power plants were introduced in great numbers in the socialist countries most of the "bugs" had been eliminated.

* Statistical sources: *Statistical Abstract of the U.S., 1972* (Washington, D.C.: Bureau of the Census, 1972), table 839, p. 516; *Survey of Current Business*, November 1973; *Rural Electrification Magazine*, October 1973; *Journal of Commerce*, 3 December 1973; *Wall Street Journal*, 3 May 1973.

Atomic Energy Safe in Socialist Lands

Because of the long period of testing and technological advance (made possible because private profits are not the goal), the socialist countries have led the way in achieving satisfactory results with rapid-breeding reactors, which increase the energy yield of uranium by about fifty times. And these, according to German Democratic Republic Prof. Max Steenbock, improve the level of thermodynamic effectiveness and thereby lower the amount of heat transferred to the environment.

Atomic energy under socialism, says Steenbock, "means a significant improvement in environmental protection—quite contrary to accompanying fears," because "all-around protective measures" and foreseeing all conceivable difficulties, are built in as integral parts of the installations. The atomic power stations of a socialist country, he claims, are "almost model examples for environmental protection," and actually give off less radioactive pollution than contained in the smoke and ash of steam power plants.*

Some environmentalists unfortunately have divorced themselves from the really important safety struggle connected with nuclear energy, the struggle for the prohibition and destruction of stockpiles of nuclear weapons. U.S. imperialism continues to build up and refine its frightful stockpiles of mass annihilation weapons, and is the main obstacle to their abolition. The ecological and other dangers associated with nuclear power plants are trivial in comparison with this.

Then there are whole new sources of energy that have hardly been tapped. These include solar power or energy from the sun, solar sea power or heat from the sea, thermonuclear or fusion energy whose main

* *Elektrie,* (GDR), No. 1, 1974.

ingredient, hydrogen, is derived from ordinary seawater, geothermal power which uses the heat from the earth's interior. Many of these are being seriously researched, however, and this research effort will certainly be multiplied. Finally, in both the United States and the Soviet Union persistent work is going ahead in the development of controlled thermonuclear power. The scientists of the two countries are already cooperating in striving to release the energy of the hydrogen atom for the use of man. We do not know how long this will take, but scientists are confident that it will succeed. And when it does there will be a virtually limitless source of energy.

Why, then, the energy crisis? Why the soaring prices, the restrictions, the shortages? The culprit is monopoly capital—monopoly capital in a new stage of development—linked up with the government to create state-monopoly capitalism. Through their use of the state the monopolies are able to create shortages. They have adapted the law of supply and demand to monopoly conditions by cutting supply, sharply increasing the price, and multiplying profits. And this technique is especially effective when dealing with everyday necessities which, in the language of economists, have a very low elasticity of demand. This simply means that people cannot afford to do without them, and keep on buying them as long as they possibly can, even under conditions of outrageous price gouging. That is what happened with meat, milk, and gasoline in 1973. And it was monopoly aided by government—state-monopoly capitalism—in every case.

Connected with the false idea that the world is running out of energy supplies is the idea that the United States is consuming "too much" energy. With 6% of the world's population, it is said, we consume one-third of the world's energy. The industrialized countries generally, it is claimed, are consuming too much, and should stop increasing the size of their economies in general.

This is a form of the doctrine of Malthus, the early nineteenth-century British economist who blamed poverty on "too many" people. Malthusian ideas are very popular with the Establishment these days: there are too many people in the United States and in other countries—especially Blacks and Chicanos—the American people "consume too much." The Establishment deplores "consumerism." All this is a very useful theory for the monopoly capitalists who certainly do not restrain their own gross overconsumption, but use these theories to justify their attacks on workers' living standards and their racist drives on the rights of Blacks.

There is excess consumption of oil and energy generally in the United States, but not by the people. The working class as a whole does not consume too much energy, but actually needs a larger supply. And even when workers consume substantial amounts, it is because of conditions created by monopoly capital.

Who Needs—Who Wastes Energy

Is the Black worker consuming too much gasoline when he drives fifty miles to work every day because he is barred from living where his job is, and because U.S. capitalism has destroyed the public transit systems? Certainly his consumption is trivial compared to that of the corporate official flitting around the country in his executive jet.

It isn't the inner-city resident, whose apartment was kept cold by the landlord even before the energy crisis, who consumes too much fuel oil, but the corporate skyscrapers that are kept alight all night and simultaneously run heat and air-conditioning power loads in the daytime. In connection with the energy crisis of 1973, much was revealed about gross waste of energy by the big corporations. The Du Pont Corporation, which can hardly be suspected of anti-industry bias, has studied the energy consumption patterns of scores of large corporations and concluded that there is an average wastage of 15% or more. The U.S. military, directly and through munitions production, wastes huge quantities of energy.

Yes, it is necessary to put U.S. energy consumption on a more rational basis, and to economize. But no cuts should be imposed on the working people of the country. Economies should be imposed on big business, the military, the wealthy. In fact, all military use of energy should be canceled out, as part of complete and general disarmament.

But what about the international aspect of the question; the fact that our share of world energy consumption is so large in comparison with our share of population? It isn't that the United States consumes too much. It is that developing countries oppressed by the imperialist empires consume too little. But the potential supply of energy is adequate for all. What the developing countries require is industrial development that would make it possible for them to consume more energy and have a higher standard of living.

It is ironic—90% of the established oil reserves and 55% of the natural gas reserves of the nonsocialist world are in the energy-starved developing countries. An end to colonialism means an end to colonial prices for the raw materials they possess. An end to economic domination by imperialism means the building of their own refineries, chemical plants, steel mills, and factories to produce consumer goods. For the corporations of imperialism this transition to political and economic independence by the developing and oppressed countries is a painful process. They have not given up the struggle. When they are forced to give up the old colonial relationships they maneuver to develop relationships based on neocolonialism.

We have charged that the monopolies have created the so-called energy crisis, and have shown that there are plentiful sources of energy in this country and in the world for now and for the indefinite future. Later we will show specifically how the monopolies and the government combined to create mass panic in the United States and in other capitalist countries with cooked-up statistics about a shortage of oil, and how they used this panic to put over the most expensive price squeeze on the consumer in all history. But the real energy crisis is the product of a number of historic processes coming together, creating a number of contradictions that demand fundamental resolutions. There is, first, the conflict between the energy needs of the people and the monopoly corporate control of the sources, production, and distribution of energy for the sole purpose of maximizing private profits. This is the conflict between the economics of corporate profits and economics of human welfare. There is also the contradiction between the growth of energy production with its more sophisticated and complicated energy technology, requiring centralized planning and control, and the inherent incompetence, the anarchistic dog-eat-dog nature of monopoly capitalism. Overall planning and monopoly capitalism are incompatible; yet without overall planning there can be no basic solution to the energy crisis. It can be relieved but not solved. Lastly, there is the conflict between the greater need for science and scientists and the decline in corporate and government money for science and the declining number of students in science.

The developing countries of the world, the source of most of the capitalist world's raw materials, are moving rapidly toward taking control and ownership of these materials, of which oil and gas are the most important. This trend has gathered momentum and is basically irreversible, and is taking place in countries with widely differing social systems.

This sharpening conflict between the countries providing the raw material and the international monopolies is another, major element of the real energy crisis for the imperialist countries and requires new and more basic solutions. Some in the United States still advocate the use of "gunboat diplomacy." Such a policy was realistic when only the imperialist countries had gunboats; now such a policy is suicide. The struggle for national independence, the breakup of the imperialist empires, is an historic process that has reached a point when it is possible to demand a basic resolution, an end to national oppression and exploitation. The energy crisis is a reflection of this historic shift.

The socialist countries, headed by the Soviet Union, the world's second largest producer of oil and gas and very largest source of other vital raw materials, set a shining example of steady, balanced, environment-conscious development of energy supplies and distribution, on the basis of true international cooperation. This is speeding the advance of the socialist countries in economic competition with capitalism, and increasing the attractiveness of socialism to the working people of the world. It represents a smashing defeat for the U.S.-dominated oil cartel, which for decades strove to starve the Soviet Union and other socialist countries of energy sources. Embargo and economic blockade are inventions of U.S. imperialism.

The gains of socialism over imperialism in the energy field are a third major element of the energy crisis of world capitalism.

The overall development within which the energy crisis takes place is the historic shift in the world balance of forces. The main result of this shift is that the imperialist, colonialist empires are no longer in a position to force their will, militarily and economically, on the world. The old solutions imposed by imperialism in the past are no longer acceptable, possible, or realistic when it deals with nations and peoples who are ready and able to fight against oppression. It cannot ignore the reality that most of the socialist countries, and in the first place the Soviet Union, back up their anti-imperialist policies by giving material and military aid to the peoples fighting imperialism.

The old imperialist instruments of blockade and embargo do not work because the underdeveloped countries can trade and receive credits from the socialist countries on a basis of friendship and equality. Imperialism is being forced to accept different kinds of solutions, solutions based on equality, on economic laws of value.

This new level in the historic process, the new power of the forces propelling the overall world revolutionary process is an important factor molding the energy crisis for the imperialist countries, demanding radical and revolutionary solutions.

part 2

• THE STRUCTURE
• THE METHOD
• THE CULPRITS

4 • Interlocking Monopolies

What makes it possible for the corporations to create and use shortages as instruments of maximizing profits? What is there about the present moment that enables them to create one of the biggest rip-offs in history?

The answers are inherent in the new stage of capitalist development. These new developments are not only possible, but if not challenged and changed, inevitable, representing as they do an intensification of the basic essence of capitalism—exploitation of the many for the profit of the few.

The answers are related to the changes in the structure and the composition of capitalism, which are now monopoly dominated. But even more, these changes in the economic structure have spawned changes in the political and governmental role and structure.

The name of the game is monopoly power—economic, political, monopoly power in the service of the big corporations and banks. A handful of financial-industrial corporate dictatorships have captured and taken over the industrial-financial, political-governmental complex. That is where U.S. capitalism is at. It is an interlocking complex that fundamentally dominates and dictates the plans for the economic and political sectors, and for the government.

The oil corporations have even greater dictatorial powers because the same small group, dominated by one family, the Rockefellers, has captured the total energy field—gas, oil, coal, and nuclear energy. The smaller corporate fish have been crushed, taken over, and devoured.

For an understanding of the crisis of shortages it is necessary to correctly assess this new stage in the monopolization of capitalism and its new level of interlocking relationships with the state.

Of course it is also necessary to keep in mind that the best laid plans of monopoly corporations, no matter how powerful, might well go astray because they are forced to operate within the confines of objective laws over which they have no control. Among these basic objective factors is that the corporation policies of exploitation are directed against the best self-interests of the people, and in the first instance against the working class, and thus stimulate and give rise to a movement of resistance and rebellion. It is also an objective law that the more the essence of corporate operations are exposed the greater will be the rebellion.

Americans are well aware that most of the country's industry is dominated by monopolies. Certainly the most powerful of all the industrial monopolies is the oil monopoly, the $100-billion behemoth, the biggest collector of superprofits, taking in more than twice that of any other industrial group.

What people are now being forced to learn is that, of all industries, the oil monopoly has the biggest say in their government, that they are no longer dealing with private monopolies. Private monopolies merge with gigantic banking groups to form superempires of finance and industry, and they enter into, control and merge with, and bring into their monopolistic schemes the government of the United States, not to mention state and local governments. What have now emerged are huge economic and financial monsters interrelated by means of interlocking directors, each having its tentacles in government bodies.

Government Dances to Oil Monopoly Tune

The economists Walter Adams and Horace M. Gray who exposed many of the devices of government-monopoly collusion, described the new stage in the title of their book, *Monopoly in America: The Government as Promoter.* Lenin had a more exact name for it—*state monopoly capitalism.* For this is not merely a collection of separate actions by the government to help monopolies, but an entire structure welding together the two components. Yes, each corporation retains its identity, and the government retains its special features, its limited independence. But the two are meshed together in every aspect of the affairs of the giant corporations, and in ways whereby the benefits all flow one way—toward the corporations—and the costs all flow another way —onto the working-class and middle-class taxpayer. The government is more than a promoter. It is also the enforcer. It freezes wages while letting the corporations raise prices. It creates the tax loopholes for the rich while raising the taxes of the people.

The government uses billions of tax dollars for research—pennies for research on health, the rest for research as a gift to the big corporations.

The government, through antilabor legislation,

court injunctions, and other repressive measures, has become the enforcer against strikes. It has also become the enforcer of racist laws and practices. In 1964 Congress passed a Civil Rights Act. Under this act Black, Chicano, and Puerto Rican workers were able to bring law suits against the corporations by showing that "both the practice and the pattern" of the corporation was racist. The workers have won a few of these cases. The corporations objected and the Nixon administration responded with steps to try to prevent this alleviation of racist discrimination. This is all done behind the smokescreen of the energy crisis, which has become the excuse for pushing the phony concept of the "work ethic." The idea is that it does not matter how much wages a worker gets, or what the working conditions are, only that he or she work hard.

Corporations are using the energy crisis to squeeze additional profits by speedup and by layoffs. There are no laws that protect workers against layoffs, and the government has become an enforcer of speedup. The U.S. Labor Department has a special subdepartment to carry out a concrete plan for speeding up production, for increasing labor productivity for each industry.

There is a shuttling of directorates between the corporations and the government. Big business executives take sabbatical leaves to run the affairs of government. And in turn, most of the bigshots who work for any length of time in the federal bureaucracy go to work as executives of the big corporation. Spiro Agnew, who had to resign because of a lifetime of corruption and pilfering, had no trouble getting executive positions in private corporations. No doubt he will give leadership in how to pilfer public monies.

Certainly no monopoly has obtained more benefits from the government than oil. Consider the millions of acres of government land and continental shelf turned over to the oil monopolies for exploitation in return for relatively trivial acquisition royalty and lease payments.

Consider the tens of billions spent on a vast highway program, combined with the suppression of other forms of transport, to make *inevitable* the predominance of the gasoline vehicle and its gross overuse within the transportation system. And consider that all of this highway system is paid for by the taxpayers (mainly through special consumer taxes on gasoline and through tolls) and none of it by the main beneficiaries, the oil monopolies.

And finally, and most importantly, consider the fact that about two-thirds of the profits of the big five of the international oil cartel come from investments abroad, a higher percentage than any other monopoly group obtains. Consider that not a nickel of these profits could be realized without the massive diplomatic and military assistance of the government, without its air and naval bases, its fleets stationed in the vicinity of every major oil field owned by the oil cartel, its armed forces stationed in the major consuming countries of Western Europe and Japan (where the U.S. monopolies continue to dominate the refining and distribution of oil).

The U.S. government spends over $100 billion a year on military, political, and economic aspects of international affairs. If we consider that more than one-third of all the profits on foreign investments accrue to the oil companies, it is fair to estimate that one-third of this government spending—or $33 billion per year—is for the benefit of the oil companies.

You Pay Their Taxes

Again, this huge bill is paid by American working people in the tax deductions from their paychecks.

Practically none of it is paid by the oil companies themselves.

Another sign of the increasing scope of state monopoly capitalism has been the steady shifting of the tax burden from the rich and the corporations to the poor and the working class. Whereas at the end of World War II the rich and the corporations paid 55% of all federal taxes, and the working class, the poor, paid 45%, now the working class and the poor pay 70% and the corporations and the capitalists only 30%.

Despite all the talk of closing tax loopholes they open wider and wider every year. But no group of corporations comes close to the almost completely tax-free status of the oil monopolies, especially the giants of the international cartel. Their particular forms of tax robbery are really outright gifts to them of billions of dollars. For instance, one loophole, the notorious "depletion allowance," is a deduction from income which in turn lowers taxes. Thus the oil corporation, in addition to a deduction of its initial investment for tax purposes, gets a continuing tax writeoff each year for no investment at all!

Another loophole is called "intangible drilling expense." It is an invitation to throw in any amount as "intangible." If you make $20,000 a year in oil investment, you deduct $4,400 as if you never received it! Because these special tax gifts are available to all investors in oil, oil has become a hiding place for all kinds of tax evasions.

Another loophole is credit for payments to foreign governments. Because of this tax gift, the little taxpayer is compelled to subsidize the leasing and purchasing arrangements made by the corporations even as they grow fat on the profits from the oil they extract from foreign countries.

These gifts are called incentives. It is like giving incentives to thieves, or calling ransom an incentive.

In 1971 Standard Oil of Ohio had a $55-million admitted profit. It paid not one penny in taxes! Besides not paying any taxes it had a reserve negative tax of some $50 million on which it can draw for nonpayment of taxes for years to come.

While the official corporate tax rate was 48%, this is what was paid by the five largest oil monopolies in 1971, in federal taxes, as a percent of profits before taxes:*

Exxon	7.7%
Texaco	2.3
Gulf	2.3
Mobil	7.4
Standard Oil, Calif.	1.6
Weighted Average	5.0

Thus, the oil companies, through their notorious tax rackets, escape almost completely from payment of taxes on their fabulous profits. Three of the largest corporations pay a rate that is one half of the rate paid by a widow of sixty-five years of age, with an income of $5,000 a year.

In fact, the total cost to the taxpayer of the special tax subsidies to the oil monopolies has been about $140 billion since these loopholes were put into law in 1926. The cost in 1971 alone was over $1.6 billion.**

This $1.6 billion alone was nearly four times the amount devoted to the government Head Start program for preschool children, twenty-one times the funds that have been allocated for the Neighborhood Youth Corps, eight times as much as was spent for on-the-job training of veterans, and eighteen times the federal support for the arts and humanities.***

* Source: Philip M. Stern, *The Rape of the Taxpayer* (New York, Random House), p. 229.
** Ibid.
*** Ibid.

But these losses to the taxpayer, enormous as they are, underestimate the true tax burden that the oil monopolies and their agents in government have foisted onto the backs of workers. This is accomplished by having a large part of the oil industry operations subsidized by the government *at the taxpayers' expense* and not accounted for in the estimates. For example, the tanker fleet owners and users have always enjoyed federal construction subsidies, loans, and mortgage insurance. And they have operated ships leased from the government at favorable terms.

The oil companies prefer to register these tankers under the flags of Liberia, Panama, and other nations where they get particularly favorable tax treatment. Standard Oil's Panama fleet, for example, is not subject to Panamanian income tax since its profits are "at sea" rather than within Panama, resulting in "virtual freedom from taxation." Shippers arrange their business so that "their profits are not liable to taxation in any country," notes a report of the Organization for European Economic Cooperation. "The abysmally low wages and poor working conditions of the nonunion seamen, permitted under these flags, was, of course, an additional attraction." Thousands of jobs for U.S. workers have been lost through these runaway operations by the fleet-footed oil thieves.

And the energy crisis is being used to revive a World War II form of government giveaway to the monopolies. Under pending congressional legislation, $10 billion to $20 billion will be spent on research and development of new forms of energy production. Of course such research is desirable. But as planned in Washington the results will all be turned over free to the fuel and power monopolies for their greater profit.

Power that the oil trust has had in the government is an old story. Our concern here is how much this power has grown, how much more arrogant the monopolies have become in their manipulation of government regulation for higher profit, for unrestrained robbery of consumers.

During the four years of this country's active participation in World War II, when there was a *real* shortage of gasoline, the average retail price went up one and a quarter cent, and there was a rationing system enabling people to get allotments of gasoline without excessive fuss. Of course there were inequities and the rich got the better of it. Of course there was oil company profiteering. But it was contained within certain bounds.

Now, with a *created* shortage of gasoline the average retail price has gone up at least twenty cents per gallon, and the end is not in sight. There is no rationing system, the supply of gasoline is doled out by the government under monopoly-dictated regulations designed to condition the consumers to line up for hours for their occasional allotment of overpriced gasoline, like trained circus animals stand on their hind legs and beg for food.

Before the cooked-up energy crisis the oil monopolies were "unhappy." Independent elements were squeezing into the picture—in production, in refining, and in distribution. From time to time the monopolies were compelled to initiate local "price wars" forcing the sale of gasoline at only modest profits. The supply-demand balance was unfavorable to the monopolies. Though retail prices of petroleum were going up it was less rapidly than other commodities and services. Profits of oil companies were going up too, but less rapidly than those of big business generally.

All their monoply power notwithstanding, the oil companies could not escape the consequences of capitalist anarchy of production. The oil shortage was devised to change this at the expense of all independent elements in the oil business. And the government, the whole apparatus of state monopoly capitalism, was enlisted to carry out this monopoly conspiracy.

The energy crisis developed simultaneously with the Watergate crisis of the Nixon administration, and the two are interrelated. Watergate was Nixon's attempt, backed by the most powerful and reactionary monopolies, to end all democratic restraints on his centralized power, to carry out the orders of monopoly power without the inconvenience of constitutional procedures, to organize secret police-state repression against all opposition. The biggest Watergate crimes were not the bugging of the Democratic National Committee and the raid on the office of Ellsberg's psychiatrist. Much more serious were the crimes in the area of the political-industrial-military-complex, the illegal invasion and secret bombing of Cambodia, the criminal genocidal bombing of Hanoi and Haiphong, the collaboration with ITT and Kennecott in organizing the bloody fascist coup in Chile. And much more serious were, and are, the crimes in the economic and social areas—the freezing of workers' wages while prices soar; the impounding of tens of millions of funds destined for social needs; the unconstitutional, illegal fascist-like attacks on the Black people—on the job front and on the education front—the murderous assault on the Indian people in the second atrocity of Wounded Knee.

The oil shortage and its handling by the Nixon administration is one of its biggest economic crimes against the American people yet.

We are paying a hundredfold for the dairy trust's donations to the Nixon slush funds. We are paying over and over for the contributions to CREEP and Rebozo by bankers, auto magnates, housing firms, metal companies, and so on. But the oil companies and oil tycoons put more money in the Nixon pockets than any other industrial groups. And the bill to the American people is correspondingly higher.

The influence of the oil monopolies on the U.S. government is not wholly a mechanical, institutional affair. It is exercised at the most decisive level by the *personal union* of the top men of the financial oligarchy with the top posts in government.

Oil and Foreign Policy

The Rockefeller family, for decades, and to this day, keeps the government running in their own oily bath, to their own maximum advantage. Nelson Rockefeller has been the front man, politically, for the Rockefeller-Standard Oil group for thirty years. During this time he has continually occupied either a key full-time government post, or has been in a decisive advisory role.

In 1942, when the Standard Oil interests were still concentrated in Latin America, Rockefeller was coordinator of inter-American affairs, and has reappeared as Latin American policy maker at every crucial interval. But as the Standard Oil grip on the Middle East, Europe, and the Far East expanded, Rockefeller's foreign policy activities correspondingly broadened. They reached a certain peak when he was made Eisenhower's foreign policy adviser in 1954 with the unusual instructions: "You are requested to attend the meetings of the Cabinet, the National Security Council, the Council on Foreign Economic Policy, and the Operations Coordinating Board."*

The very term "energy czar" is conceived in the spirit of Watergate. It signifies the appointment of an individual with arbitrary power over a vital part of the

* *New York Times*, 17 December 1954.

country's life, subject to absolutely no democratic restraints. But this czar isn't selected by inheritance, by divine right. He is selected by the oil monopolies, and he isn't exactly a czar. Satrap of the czars, he is the one who hands down and enforces their merciless edicts against the people. The identity of the real czars is well enough known. They are the Rockefellers, who personally contributed an identified $286,700 to Nixon and whose Exxon and Standard Oil of California executives, alone, contributed another $293,747. They are the Mellon bunch, whose top member Richard Scaife contributed $1,003,000. These tycoons bought the right to select the satrap to put over the massive oil plundering of the American people. That right flowed not only, or even mainly, from their contributions to Nixon, but from their supreme power position in the financial-industrial oligarchy of the United States. And in this sense the Rockefellers continue to occupy the leading position. In short, Rockefeller was appointed to all of the main strategy and operation centers of the government. The corporate godfather was given the key that opened the doors to all government operations.

Rockefeller rivaled Nixon for the Republican nomination in 1958 and lost, which didn't hinder his ability to dictate U.S. foreign policy. Henry Kissinger spent his entire adult life as top foreign policy and military policy aide to Nelson Rockefeller and the Rockefeller brothers, and retained these close ties after becoming Nixon's top foreign policy man. In addition, according to press reports, Rockefeller personally was consulted by Nixon at least once a month and had frequent telephone contacts.

When Nixon, in the latter half of 1973, became absorbed in defending his political life from the Watergate scandals, Kissinger advanced to the joint posts of secretary of state and foreign policy adviser to the President. Moreover, Kissinger let the press know that he, and not Nixon, was really running foreign affairs. He was referred to widely by the press as "President for foreign affairs." He claimed to have organized the adventurous armed forces alert of October 1973 on his own responsibility, and merely informed Nixon of it when it was all prepared.

But Kissinger's power flows not from his admiration of Bismarck, nor his supposed genius. It flows from his representation in Washington of the Rockefeller-Chase Manhattan-Standard Oil interests.

The removal of Nixon is an important and necessary step to save American democracy, to prepare for a counteroffensive against monopoly superpower. But this step by itself will not remove the evils and dangers of state monopoly capitalism, nor the sinister power of the petroleum monopolies in Washington.

5 • The Rockefellers and the Oil Trust

While the oil corporations are a model for the study of monopoly state capitalism, the Rockefeller dynasty is a perfect specimen for the study of plunder, looting, extortion, corruption, conniving grants, total lack of social responsibility, murder for private profits, antiunion frame-ups, racism, trampling on human rights, and a total disregard for the Constitution when it is a hindrance to lining the pockets of the Rockefellers.

The Rockefellers are in the very center of the energy crisis. Their riches are riches from oil. They are the first family of the oil pirates of the world. More than anywhere else the roots of the energy crisis are in the Rockefeller oil empire.

The Rockefeller-Standard Oil monopoly is one

hundred years old, the oldest and most powerful monopoly of modern capitalism. The Rockefellers are great advocates of "independent," "free" enterprise, but they did more to annihilate independent capitalism than any other group in America. *Life* magazine wrote of the establishment of the Standard Oil Company of New Jersey in 1899: "The new setup was established by John D., Sr., after maneuvers so ruthless that they made him one of the most hated men in America."* To call Rockefeller's action "maneuvers" is to cover up for murder, deception, and corruption. No war between gangsters has ever matched the brutality of the Rockefeller war for oil.

In the first stage of their monopoly the Rockefellers obtained a practical monopoly of petroleum refining in the United States by forcing out of business virtually all rivals. Using the millions thus accumulated, they stole and bought up oil lands and obtained a leading position in the extraction of crude oil also. They used massive political bribery of government officials to get approval for their pipelines and to block the pipelines of rivals. They were totally involved in the earlier Watergates. They conducted local price wars to drive small producers and distributors out of the market, a practice they continue to this day—except of course in periods of acute shortage when they operate mainly by depriving rivals of supplies of oil.

Rockefeller's first foray in plunder was at the expense of the armies of working men during the Civil War. Here John D., a common trader of merchandise, made enormous sums from the soaring wartime prices. Shortly thereafter, in 1865, following the discovery of oil in the northwestern corner of Pennsylvania, Rockefeller was persuaded by a sharp refinery technologist named Andrews to throw his loot into

* *Life* Magazine, 23 January 1950.

the oil business. Rockefeller took charge of the bookkeeping side of the venture, perfecting his pennypinching, vicious antilabor techniques. His company soon had the biggest oil refinery in Cleveland.

Capitalism engenders bestial qualities in certain people. Those who developed these qualities can become good robbers. John D. "Reckafellow," aptly named by fellow robber baron Andrew Carnegie, was a fine example.

"I cheat my boys every time I get a chance. . . . I trade with the boys and skin 'em, and I just beat 'em every time I can," was the golden rule adopted by John D.'s father and followed by his son*.

Railroads possessed the strategic power of life or death over the oil companies who, to survive, must ship their oil to the refineries and markets. Relying on his superior oil supply as a bargaining weapon, Rockefeller was able to "skin" the railroads into paying his company secret freight rebates. They also turned over to him fifteen cents of every forty cents they charged others for transporting crude oil from the Pennsylvania oil fields. And for the delivery of the refined product to the market he actually paid forty cents less than his rivals. In addition, the railroads engaged in espionage on his behalf, regularly reporting whom his rivals were selling their products to, and the quantities and prices involved.

Secrecy and conspiracy—Watergate style—has been the Rockefellers' stock-in-trade from the earliest beginnings. It is a hallmark of monopoly. The 1882 Rockefeller Standard Oil trust—the behind-the-scenes shift of the capital of many companies to nine "trustees"—was formed in exactly this manner.

Rockefeller gave his weaker competitors a single proclamation—either sell to him at his own price or

* Allan Nevins, *John D. Rockefeller* (Krause Reprint, 1940); cited in Robert Engler, *The Politics of Oil*, (New York: The Macmillan Co., 1961), p. 478.

be ruined. Determined to control the oil business of the world, he persuaded his stronger rivals to join him in his plan while wiping out others. He systematically gobbled up gas and electric companies, interurban transport lines, copper concerns, iron ore mines, and railroads all over the country. Rockefeller, the grand master of oil, discovered he could dominate the oil world by controlling the railroads, pipelines, and refineries.

In 1911, at the time of the "slap-on-the-wrist" antitrust action against Rockefeller, his trust handled about 90% of the nation's oil business. In 1879, on an invested capital of $3.5 million, dividends of $3.15 million were paid—a 90% to 100% rate of profit was the norm.

Essential to Rockefeller's grand designs were the banks that could provide the financial wherewithal for the endless schemes and intrigues directed at ownership and control of all major sectors of the economy. Beginning at the turn of the century, in alliance with James Stillman, president of the then National City Bank, and continuing with the penetration into a whole series of banks (of which Chase Manhattan has been the most noteworthy), the money capital flowing out of bank domination became a mighty ramrod by which the Rockefellers could crush all rivals.

Ludlow to My Lai to Attica

The Rockefellers are among the most notorious antilabor barons of industry. Besides dominating oil they were also interested in rivaling the Morgans in coal and steel and gained control of the Colorado Fuel and Iron Company. In 1914 the coal miners at Ludlow went on strike. The miners, members of the United Mineworkers of America, were evicted from their company-owned homes in midwinter and forced to live in tent colonies surrounded by National Guardsmen. On Easter night, while the miners and their families were sleeping, the National Guard attacked and set fire to the tents and machine-gunned the fleeing people. One man, six women, and thirteen children were killed, dozens were wounded. John D. Rockefeller, Jr., personally took charge of the strikebreaking and justified it as part of a "principled fight" against trade unionism.*

In 1892, at the Carnegie steel works known as Homestead, the company president, Henry Frick, imposed a mammoth wage cut after failing three years earlier to bar the union. This time he hired a battalion of armed Pinkerton thugs to take over the workers' jobs. The strike began as the plant shut down and Frick brought out the goons. In heroic struggle the workers held off the company onslaught for five months until finally they were overcome by government troops. With the aid of an anarchist attempt to assassinate Frick the steelmasters won the day.

From his far away European retreat Carnegie wrote his class brother John D. Rockefeller: "Congratulations all around—life worth living again—how pretty Italia." And to this John D. enthusiastically replied, "approving his course and expressing sympathy."**

Half a century later his grandson, Nelson Rockefeller, personally gave complete approval to the war waged by U.S. imperialism in Vietnam, a war conducted mainly with Ludlow-type genocidal tactics against the Vietnamese people, including the massive use of napalm, manufactured out of Rockefeller-produced gasoline. The bloody trail of the Rockefel-

* Boyer and Morais, *Labor's Untold Story* (New York, Cameron Associates), p. 190

** Matthew Josephson, *The Robber Barons: The Great American Capitalists, 1861-1901* (New York: Harcourt, Brace & World, Inc.), p. 372.

lers goes from Ludlow to My Lai to Attica. In Ludlow the murder was to get cheap coal. In Vietnam, the mass murder was for cheap colonial-priced oil.

The Rockefellers went to great lengths to keep unions out of the oil industry. Strikes in New York refineries during World War I "were broken savagely by Standard's guards and the police and deputies of Bayonne. Following so closely after the 'Ludlow massacre' of the wives and children of striking Rockefeller coal miners by the Colorado militia, these uprisings stirred liberal labor partisans throughout the country and plunged the Rockefeller name to its nadir."*

Thereafter the Rockefellers introduced company unions and with their aid were able to prevent unionization of the industry for about fifteen years longer than other basic industries. But finally in the 1950s the oil workers broke the Rockefeller-Mellon resistance and established industrial unionism in much of the production and refining of crude oil.

Racism: A Rockefeller Heritage

Wherever there is an attack on the working class there is the use of racism as a weapon against it; the Rockefellers have a long history of racism. As late as 1950 only 0.7% of the workers in the crude petroleum and natural gas industry were Blacks, as compared with 6.0% in all other mineral industries combined. And this despite the fact that the largest center of oil production was in east Texas and Louisiana with their large Black populations. The Rockefeller-dominated Metropolitan Life Insurance Company attained notoriety with the brazen segregation policies it followed in enormous housing developments, including the

* Harvey O'Connor, *The Empire of Oil* (London, J. Calder), p. 167

eviction of white tenants who tried to help Blacks obtain entry into all-white Stuyvesant Town development in New York City.

Only after years of mass struggle by the Black people, and enactment of civil rights laws, and court battles, did the Rockefeller banks, insurance companies, and oil companies turn to the hiring of Blacks for other than the most menial occupations.

Discrimination against the Chicano people in the southwest is also part of the Rockefeller tradition. Chicanos are often employed in the most dangerous jobs on the drilling rigs, while being paid low wages and denied any employment security.

The Rockefellers and the other oil tycoons stole many thousands of square miles of Indian lands in Oklahoma, Wyoming, and other states for the oil under the ground. The much-publicized payments to the Oklahoma Indians were a drop in the bucket compared with the billions of dollars' worth of oil taken out of their land by the oil monopolies. The robbery still goes on. The per capita income of Oklahoma Indians in 1970 was only $1,614, less than half the per capita income of white people in the United States.

The oil companies have engineered a similar robbery of the Alaskan Eskimos and Indians. The U.S. government allotted a total of a billion dollars "for the benefit of" the Eskimos and Indians. As is being done in Oklahoma, this will be doled out to them over a twenty-year period, and under the control of the U.S. and Alaskan state authorities. Many a government bureaucrat will have his hand in the till. In that same period the oil monopolies expect to get something like 30 billion barrels of oil worth $200 billion from the lands of the Eskimos and Indians. The payment to the native peoples amounts to *one-half of one percent* of the values to be taken.

The robbery of their oil will also cost the Eskimos and Indians the destruction of their hunting and fishing lands, the disruption of their way of life and culture, and the damage to their environment by the pipeline and other operations.

No group has surpassed the Rockefellers and the other oil tycoons in robbing the pubic purse, both through public expenditures on their behalf and through tax loopholes. At the same time the government has spent literally hundreds of billions of dollars in overseas military outlays to help the oil companies gain and keep control of the oil of Latin America, the Middle East, and the Far East, and to create conditions enabling them to dominate the refining and distribution of petroleum in Europe and Japan. Thus, the Truman Doctrine, with its bases in Greece, Turkey and Saudi Arabia, had as a principal objective the domination of the Middle East oil area. And the largest beneficiaries of the Marshall Plan, which opened up Western Europe to U.S. investment on a huge scale, were the Rockefellers and the oil industry.

The oil firms, as well as the auto industry, have profited from the government's one-sided transportation policy featuring highways at the expense of mass transit systems and railroads.

Today, the oil industry has expanded to include other significant groups beside the Rockefellers. But if their monopoly is no longer total, it must be recognized that the industry they still dominate is immeasurably bigger than when they held nearly 100% control. Moreover, their grip encompasses many areas besides oil.

The Rockefellers are among the four superbillionaire families in the United States, with fortunes in the mid-1960s estimated between \$5 billion and \$10 billion. However, their control over industry and finance is greater than that of any other family.

They control the Chase Manhattan Bank, the Metropolitan Life Insurance Company and control or share control in assorted industrial, transportation, financial, real estate, and munitions companies. And above all, they control three of the world's largest oil companies, Exxon, Standard Oil of California and Mobil, besides such lesser oil giants as Standard of Indiana and Marathon Oil. All these were part of the original Standard Oil trust, which was nominally but not actually broken up after World War I following an antitrust suit.

In forming the original Standard Oil combine, Rockefeller bankrupted his weaker rivals, and bought off his stronger rivals, bringing them in as partners in his company. The descendants of Rogers, Brewster, Pratt, and other Rockefeller partners continue to buttress the Rockefeller power structure, in economics as well as in politics.

The total assets controlled by the Rockefellers now easily exceed $100 billion, representing an economic empire larger than the entire economies of all but a dozen or so of the world's countries.

On the international scene, the Rockefellers have followed the same tactics as on the domestic front. They formed a cartel with the other world oil giants that they could not annihilate—Mellon's Gulf Oil, Texaco, controlled by Chicago financiers and the Manufacturers Hanover Bank, the Royal Dutch Shell group, controlled by the Dutch and British monopolies, and British Petroleum. But even within the cartel, their constant aim is to expand their share of oil and profits. In the Middle East, they have elbowed out their partners so that now the U.S. companies have twice as much as the British and Dutch, and Standard Oil companies along with their close associate, Texaco, completely own the richest oil fields in the world, those of Saudi Arabia—or did until the Saudi Arabians began to cut down their share in 1973 prior to taking over the fields completely. Exxon affiliates are also the leading forces in the production of oil in Venezuela and Canada, the two largest producers in the Western Hemisphere outside of the United States.

Corresponding to their financial power, the Rockefellers have had a leading part, and often a dominant part, in running the politics of the United States and a number of other countries, The Rockefellers virtually owned Presidents McKinley, Taft, and Eisenhower. Eisenhower with Nixon's direct participation turned over the State Department and the Commerce Department to the Rockefellers. They have held the post of secretary of state—or chief foreign policy adviser-continuously for more than twenty years.

The third generation of Rockefellers, the Rockefeller brothers, continue the family tradition. Nelson Rockefeller, as governor of New York state, followed a "tough" policy toward the poor and oppressed, highlighted by his responsibility for the Attica massacre.

The basic foreign policy line of Standard Oil and the Rockefellers was set forth by Standard Oil (New Jersey) treasurer, Leo D. Welch, in 1946: "As the largest producer, the largest source of capital, and the biggest contributor to the global mechanism, we must set the pace and assume the responsibility of the majority stockholder in *this corporation known as the world*. Nor is this for a given term of office. *This is a permanent obligation.*" (Emphasis mine—GH).

The Rockefellers personally organized the big military buildup of the 1960s, and created the policy line that set the stage for U.S. aggression in Vietnam and U.S.-backed Israeli aggression against Arab lands.

This was in the "Rockefeller Reports," issued in 1958, dealing with military, foreign, and domestic policy. The report on "International Security: The Military Aspect" is a hard-line concoction of anti-Soviet hysteria. It denounces the concept of peaceful coexistence and flatly advocates the use of nuclear weapons: "*The willingness to engage in nuclear war, when necessary, is part of the price of our freedom.*"(Emphasis mine—GH). It establishes a readiness to attack any country whose policies the Rockefellers decide to label as "subversion" or "concealed foreign penetration." It calls for a $3-billion-per-year increase in the military budget, a proposal that was carried out by the next administration—Kennedy's.

Rockefellers' "Boy," Kissinger

The project director of that Rockefeller special study was Henry A. Kissinger. It was he who drafted the report; it was the Rockefellers who decided the policy.

The Rockefeller studies were conceived as a response to the first Soviet sputnik, a pioneering scientific achievement of great potential benefit to all humanity. But the Rockefellers and their ilk could see only its military significance; that it meant the end of what they had considered their strategic dominance over the Soviet Union and the world. And it was in these reports that the Rockefellers first popularized the phrase "two superpowers" tarring the Soviet Union with their own imperialist brush, the phrase which has now been picked up by the Maoists and other anti-Sovieteers here and abroad.

This same Kissinger, as Nixon's foreign policy chief and secretary of state, has been forced to participate in formally accepting with the Soviet Union the principle of peaceful coexistence and an agreement to prevent the outbreak of nuclear war.

But the cold war ideology of Kissinger and Rockefeller has not changed one bit. Rockefeller gave complete and continuous support to U.S. agression in Vietnam, and to Israeli aggression in the Middle East. Kissinger, in his 1974 policy speech in London addressed to the European NATO members, attacked the "dangerous" situation "in which the public mind identifies foreign policy success increasingly with relations with adversaries, while relations with allies seem to be characterized by bickering and drift." He called on other capitalist countries to join with the United States in a united program to solve the energy shortage, not in cooperation with but in opposition to, the socialist countries. "The energy crisis of 1973 should become the economic equivalent of the Sputnik challenge of 1957."

Kissinger's speech was an indirect appeal to the European countries—you join an alliance directed against the Soviet Union and the United States will give up the policies that lead toward detente.

Thus, having been forced, against their will, by powerful world currents to move toward detente, toward peaceful coexistence, the Rockefellers and their oil monopoly associates see their empire slipping away into the hands of the rightful owners of the oil, the people of the lands where the oil is located. And they are hoping through the restoration of tension with the USSR, through reviving the cold war mentality, through tightening up the collapsing aggressive NATO block to create conditions for holding on to every drop of the capitalist world's oil. They want to maintain and extend their position as "leading stockholders of this corporation known as the world," and to that end are willing to impose endless sacrifices on the public and risk the supreme danger of nuclear war.

Now Nelson Rockefeller has set up a new Commission on Critical Choices for America. Press comment stresses the commission as a publicity buildup for Rockefeller's presidential bid, as was charged about the earlier reports. Doubtless Rockefeller has this in mind, but that is not its main objective. Now, as fifteen years ago, the Rockefellers are anxious to put out a highly publicized, prestigious piece of propaganda for *their* choices for America. One can be certain that their choices will be consistent with the predatory aims of the oil trust and of U.S. imperialism generally. The time is far advanced. The national liberation movements and the socialist camp have made giant strides since the first Rockefeller reports. The rot of a decaying capitalism has penetrated much deeper, especially in the United States.

The Rockefellers are more desperate than before. There is more danger of adventurist proposals with catastrophic results.The forthcoming Rockefeller reports must be very critically examined by all antimonopoly and antiwar forces; and if they adhere to historic Rockefeller aims, they must be broadly exposed.

6 • The Seven Thieves and Oil Imperialism

Oil and imperialism are inseparable. Oil has been a major factor in molding and structuring U.S. capitalism, and the most important single material object of modern imperialism.

In the twentieth century the American, British, French, and German monopolies entered the struggle for the oil of the Middle East, having already seized concessionary rights from feudal governments under pressure of their fleets, and partial armed occupation.

The World War I battles of Lawrence of Arabia and the other pirates of imperialism were fought for oil. German and British imperialists, along with Russian capitalists, divided the oil of Baku and Grozny. The Russian Revolution deprived the German imperialists of Caucasian oil and played a major role in weakening the German armed forces. Following the Revolution, the British invaded the Caucasus in an attempt to seize not only Caucasian oil, but the promising reserves in the Urals and Siberia. Meanwhile, the Dutch tycoons seized the oil in Indonesia.

In 1914 the U.S. armed forces invaded Mexico to force a change of government and gain concessions for U.S. oil companies at the expense of British rivals. Following World War I, the United States emerged stronger than its exhausted allies and demanded and got a substantial share of the Middle Eastern oil booty. Standard Oil and other American companies penetrated into the oil fields of Romania and Hungary, as did British and other companies. U.S. imperialism, backing the bloody dictator Gomez, won out in the struggle for Venezuelan oil in behalf of Standard Oil, while Gulf Oil got the Colombian prize. U.S. companies also obtained positions in Canadian oil.

Again, in the Second World War, the Japanese imperialists fought for the oil of Indonesia, while the German tank armies of Rommel drove across Libya and Egypt heading for the oil fields of the Middle East. The Nazis threw everything into their drive for Caucasian oil, but reached only its outskirts and were driven out by the heroic Soviet army before they could extract a single barrel. The gallant defenders of Stalingrad shattered the oily dreams of German imperialism forever. They struck the single most powerful blow that changed the course of history.

This time U.S. imperialisim, as the only real victor among the imperialist powers, seized the lion's share of the capitalist world's oil for Standard Oil, Gulf, and Texaco—for the Rockefellers. In a radical redivision of Middle Eastern oil the leading position was transferred from the British and Dutch to the U.S. corporations.

But the struggle among the imperialists continues. The Japanese and West German imperialists, seemingly smashed durng World War II, have returned to the world oil scene and are obtaining fresh concessions in oil producing countries.

Between the two world wars the main oil monopolies organized a world cartel that divided up markets and concessions. They set colonial prices and used their combined economic power and the military and political force of their governments to plunder, loot, and keep control over oil producing countries as their private preserves.

Seven Sisters / Seven Thieves

The members of the world cartel are the so-called Seven Sisters (a term used in their literature which hides their imperialist essence). The Seven Thieves is more appropriate. The most powerful world monopoly of capitalism, they are: Exxon Corporation, Texaco Inc., Gulf Oil Corporation, Mobil Oil Corporation, Standard Oil Company of California, Royal Dutch Shell Group, and British Petroleum Company—the first five being wholly U.S. concerns which received 80% of the profits in 1972.* While there are rivalries among them and periodic redivisions of the spoils, they generally act as a coherent, single monopoly in the economics and politics of world oil.

Five of these seven are headquartered in the U.S. and three of them belong solidly to the Rockefeller group—Exxon, Mobil, and Standard Oil of California. Gulf Oil belongs to the billionaire Mellon family of Pittsburgh. Texaco is controlled by a combination of Chicago interests, the Manufacturers Hanover Trust, and Texas capital. Royal Dutch Shell (the strongest partner of Standard Oil) is run from Amsterdam and London, a combination of Dutch and British interests (the latter including prominently the Rothschild

* See *Standard and Poor's Industry Surveys*, 5 July 1973, p. 89.

group). British Petroleum Company, whose stations now appear in the United States under the BP label, is owned by British capitalists and has joined forces with the Rockefeller combine as a division of Standard Oil of Ohio.

In modern times the main material basis of the "special relationship" between British and U.S. imperialism is the alliance of their oil monopolies. And the special slavishness of Dutch imperialism in relation to NATO and all the aggressive policies of U.S. imperialism is connected with the tie-up of the oil monopolies. The loot of international oil contributes much to the relative prosperity of that small country.

The role of these Seven Thieves in the control of world oil is staggering. In 1972 they owned 46% of the world's oil production and 57% of that of the capitalist world. Moreover, they effectively *controlled* virtually the entire production of the United States, raising their share of effective ownership or control of capitalist world oil to nearly 75%. In recent years the Seven Thieves have been feverishly expanding their production of oil in old concessions, and developing promising new concessions. Their total production increased 2.4 times between 1962 and 1972; their share of world production increased as well. Their domination in refining and distribution of petroleum is approximately as decisive as in production.

Until the turn of the century, the strength of the oil trust was oil from the home market. With the onset of the imperialist stage of U.S. capitalist development, the bulk of the oil production was shifted abroad. In 1930 U.S. domestic oil production accounted for about two-thirds of the world's total, but in 1960 it was about one-third.* The shift in oil production was accomplished through superexploitation.

* Harvey O'Connor, *World Crisis in Oil* (London: Elek Books, Ltd., 1962), p. 19, Cf. Research Bureau, *1973 Commodity Year Book*, p. 257.

The Seven Thieves control *60% of the giant oil fields of the Western world and 80% of the proven reserves*. The five U.S. Thieves produce almost half of the refinery production in the Western world outside of the United States. About two-thirds of their production and almost one-half of their refinery capacity are overseas.

Exxon, the Biggest Thief

The enormous economic strength of these oil companies is mightier than whole countries. For example, in 1965 alone the sales of Exxon of $11.5 billion were larger than the individual gross national product of Norway, Iran, Israel, South Africa, and Denmark, to name but a few.

The international oil companies as a group are by far the most important single private concentration of economic power within their respective countries.

Rockefeller's Exxon is the giant among giants. Not only is it the biggest supplier of oil in the United States but it also has a very large block of coal reserves, holds major uranium deposits, and manufactures nuclear fuel. It is many times the size of the original Standard Oil trust. It is, in its own words, "an organization that operates in nearly every currency, in nearly every language and under a bewildering complexity of economic and social and political circumstances."*

Strictly a holding company that does nothing but manipulate the stock of its subsidiaries, Exxon's operations extend to some thirty-eight countries all over the world, especially the Middle East and Africa which supplied half of the company's net crude pe-

* Cited in Robert Engler, *The Politics of Oil* (New York: The Macmillan Co., 1961), pp. 38-39.

troleum and natural gas in 1972. Over two-thirds of its reserves of crude oil and liquid natural gas are in the Middle East and Africa, along with 43% of its total capital.* In 1973, Exxon beat out General Motors to establish the all-time profit record for a single industrial corporation . . . $2,440,000,000—nearly 2½ billion dollars. The Big Five of U.S. petroleum—the five U.S.-owned members of the International Cartel, made a total of $6.2 billion, or nearly two-thirds of the total profits of all U.S.-owned oil producing and refining companies.

Sales of these five equalled $50.2 billion in 1972 or 37.6% of the sales of the top ten industrial corporations in the United States—$133.4 billion; and among the "Top 500" industrial corporations Exxon ranked number two, Mobil number seven, Texaco number eight, Gulf number eleven, and Standard Oil of California number twelve. In terms of assets the five ranked even higher—one, three, six, seven, and nine. Six of the top fifteen U.S. industrial corporations were oil companies.

In 1972 Royal Dutch Shell and British Petroleum were ranked one and four respectively in the *Fortune* Magazine list of three hundred largest industrial corporations outside the United States.**

In the United states the petroleum industry is dominated firstly by the five Thieves, and secondly by thirteen other large monopolies which are "vertically integrated," that is, they operate in all phases of the oil business—exploration, production, transportation, refining, distribution, and marketing.

These eighteen are known as the majors (although the term is frequently used to refer to the top ten oil companies), and they produce about 70% of the domestic crude oil, control some 80% of the refinery

* *Moody's Industrial Manual*, 1973, pp. 755, 760, 788.
** *Fortune* Magazine, September 1973, p. 204.

capacity, and market about 72% of the gasoline sold in this country. The eight largest companies have about 58% of the nation's refining capacity. The fifteen largest have about 78% , and the top twenty-one have about 85% of the total.

The largest of the majors dominate the production of natural gas. These two fuels, oil and gas, provide three-fourths of the energy used in the United States.

The eighteen top U.S. oil companies have some 460 interlocks including 132 at banks, 31 with insurance companies, 12 with utilities, 46 with schools, 15 in transportation, and 224 with large manufacturing and distributing companies, according to a confidential report prepared for the Center for Science in the Public Interest, which was quoted in the *N.Y. Post* of December 22, 1973:

> Oil company directors . . . form a cozy and exclusive club, able to take common, if not conspiratorial action . . . Outsiders simply do not know what goes on behind the closed door.

The oil industry is the nation's wealthiest. It has about twice as much admitted profits as the next largest industry. In the third quarter of 1973 admitted profits from petroleum products and refining were 26.2% of the total profits in manufacturing—*over one-fourth*. While profits increased 35% over the second quarter for manufacturing as a whole, for petroleum products and refining the increase was 61%.*

Twenty-seven percent of the land area of the United States is under lease to the oil industry or is already in production. Thirty-two states produce oil and gas. Domestically the largest oil producers are Texas and Louisiana, accounting for 63.6% of total U.S. crude oil production. But out of the fifty states in the nation Texas and Louisiana ranked fortieth and forty-first in percent of children completing high school in

* First National City Bank, *Monthly Economic Letter*, November 1973, p. 9.

1970(!)—an illustration of how the people suffer while the oil robbers wax fat.*

With more and more of the world's oil moving across oceans, the pace in building and acquiring tankers has been furious, especially since the closing of the Suez Canal by the Israeli occupiers in 1967. The world tanker fleet more than doubled between 1965 and 1972. Shipyards are filled to capacity with tanker orders. The Seven Thieves controlled 106,415,000 deadweight tons of tankers as of the end of 1972, or 55% of the world total of 192,894,000 deadweight tons. Another 10% or so were owned or chartered by lesser U.S. or British companies, and by others under their influence, making an effective control of perhaps two-thirds of the world's tanker fleets.

The Seven Thieves have consistently demonstrated the oppressive and plundering character of imperialism. They have sought out for the extraction of oil either outright colonies or semicolonies with the world's most backward and reactionary governments, and where necessary have installed such governments with the military aid of their own imperialist armed forces and other means of intervention. Neocolonialism is the creation of the oil imperialists.

Indeed it is symbolic of the politics of world oil that their largest single oil source is Saudi Arabia, which formally abolished human slavery less than a decade ago.

Nowhere has the contrast between the poverty of the masses of producing countries and the incredible wealth taken out by the owning corporations been more extreme than in the oil countries dominated by the Seven Thieves. In the early 1950's, when Venezuela was the second largest oil producer in the world, mainly controlled by the Rockefeller interests, its population was starving on an average consump-

* U.S. Department of Commerce, *Statistical Abstract*, 1973, p. 117.

tion of 1,500 calories daily, and its domestic agricultural economy had been badly weakened by the impact of the oil economy.

Nowhere did the oil corporations train local people as engineers and managers. In fact they went to great lengths to keep all local workers in subservient positions, to prevent them from gaining the technical knowledge necessary to run the oil installations.

Nowhere did the oil corporations contribute anything to the industrial and cultural development of the producing countries. Indeed, they did everything they could to hamper both. The oil of Venezuela was not refined in that country, but on nearby Dutch islands beyond the reach of the Venezuelan people. Some of the oil of Saudi Arabia (since World War II, totally within the Rockefeller empire) was refined on Bahrein Island, a British colony, but none on Saudi Arabian territory. The pipelines run from the oil fields to the ocean terminals.

The monetary payments to the workers and governments of producing countries were kept at trivial levels as long as possible. In Venezuela such payments are in the thirty-to-forty-cents-per barrel range, or under a penny a gallon. In the extreme, in Saudi Arabia and other Middle Eastern countries, total production costs were around ten cents per barrel (and still are). That is about a quarter of a cent per gallon (a penny for four gallons). That is piracy. In 1950 payments to governments of Middle Eastern countries averaged only thirty-one and a half cents per barrel, or about three-quarters of a cent per gallon. To put that in perspective—at the time the price of gasoline to consumers was twenty-seven cents per gallon in the United States, and two to three times that amount in Western Europe and Japan, the oil producing countries were getting a penny for four gallons.

Naturally these conditions led to fabulous superprofits. And these profits have soared, along with the multiplication of the output of petroleum. Reported profits of U.S. oil companies on foreign investments increased from $627 million in 1950 to $4,552 million in 1972. About half of all profits remitted from foreign investments are those of the oil giants, and mainly by the U.S.-owned five of the Seven Thieves. Preliminary data make it clear that their foreign investment profits in 1973 increased by at least another billion dollars.

The brightest jewel in the Standard Oil-Rockefeller empire is in the Middle East. Remitted profits of the oil companies from the Middle East, excluding Libya, came to nearly $2.5 billion in 1972, or almost one-third of the corresponding figure for all corporations of all industries operating in foreign countries. The rate of return on capital invested in Middle Eastern oil is beyond belief: 167% of the book value of investment. This is according to the figures of the oil companies themselves.*

For many years the superprofits from foreign oil, more than any other commodity, made it possible for the British and Dutch capitalists to import more goods than they exported. And these superprofits helped U.S. imperialism finance its global military activities, largely to promote the further expansion of the oil monopolies.

During the past decade almost all expansion in production and profits of the oil monopolies has been in the Middle East and Africa. By 1972, 70% of their total output of crude was in the Eastern hemisphere, the areas of greatest activity and most acute struggles of the global national liberation movement.

Fancy the explosive potential of the 8 billion barrels of oil pumped out of Middle Eastern wells in 1973. The political struggle over that oil is incomparably

* U.S. Department of Commerce, *U.S. Business Investments in Foreign Countries, 1960* (Washington, D.C.: Dept. of Commerce): *Survey of Current Business*, September 1973.

more explosive and signals danger to all peoples on our globe.

Israel and Oil: Monopolies Choose

"The war has shown that now even the conservative Arab regimes will try to force the United States to choose between Israel and oil."* The choice between Israel and oil has never been an option for U.S. foreign policy in the Middle East. That foreign policy has always been an Exxon-Mobil-Texaco policy, a policy of maximizing corporate profits.

Before the Second World War the Middle East oil was totally dominated by U.S.-Dutch-British monopolies. After the war two significant changes took place. The Middle East oil producing countries began the movement toward political independence. Until this point the oil monopolies had taken the oil at almost "cost of extraction." The new governments began to demand a price for the oil, in addition to the cost of extraction. About the same time, with the active participation of the United Nations, and in the first place the Soviet Union, Israel was born as a new state.

U.S. imperialism never debated between oil and Israel; but simply the best means to obtain cheap colonial oil. The United States hesitated and held back for some time before backing the United Nations resolution to establish Israel. From that moment on its policy in the Middle East has involved how to use Israel to continue getting oil at the cost of extraction. This coincided with the Zionist policy of expanding the borders of Israel. It is these interlocking interests of Zionist expansion and U.S. aggression for oil that

* Krasner, *Foreign Policy*, No. 13, 1974.

have propelled Israel's policies of aggression against Egypt, Syria, and the other Arab countries. The million Palestinian refugees from what is now Israel are victims of oil imperialism.

The October 1973 war changed this situation. The balance of forces has shifted. There has also been some shift in U.S. Middle East policy. The aim remains cheap colonial oil, but the tactics are changing.

What has become apparent to the oil monopolies is that Israel today is not the effective instrument for getting the Middle East oil on their terms. The first open signal for this change of policy interestingly enough did not come from the State Department but from the president of Standard Oil of California:

> All of this is occurring at a time when the Arab states—because of their vast reserves of crude oil—are becoming increasingly important to the future welfare of the Western world. The Arab states—and Iran—hold the key to the energy resources which fuel the industrialized nations of Western Europe and Japan. They represent the only major source to which the United States can look for any substantial increase in its crude oil imports to meet our needs.
>
> It is in the best interest of all of us who are citizens of the United States to urge our Government to work toward conditions of peace and stability. We must acknowledge the legitimate interests of all the peoples of the Middle East and help them to achieve security and a dependable economic future.*

Under the guidance of the Zionist movement, the leaders of Israel made a historic miscalculation. By relying on the support of the United States while continuing its policy of aggression Israel became totally isolated in the world and is at the mercy of Washington, D.C.

The shift in U.S. policy of course is a maneuver, an attempt at a balancing act. Juggler Kissinger shuttled

* O. N. Miller, Chairman of the Board, Standard Oil of California, letter to stockholders, 26 July 1973.

back and forth between Tel Aviv and Cairo and Damascus. The fact that the United States was in a position to dictate to Israel was obviously a selling point for him. Each mile that Israel is forced to withdraw will be used by Kissinger and the oil monopolies to solidify their position with the leaders of the Arab oil producing countries.

Neither the people of Israel nor the people of the Arab countries—nor the people of the United States for that matter—have ever been of concern in determining the policies of imperialist aggression. U.S. foreign policy is an oil policy. The aim is oil and the superprofits from oil obtained at the cost of extraction.

Mr. Akins, U.S. ambassador to Saudi Arabia, places the question quite well: The critical issue, he proposes, is whether the oil producing countries can maintain an oligopoly without the help of the Seven Thieves. Mr. Akins grudgingly admitted that this is a definite possibility. In fact, the answer is obvious. The process of nationalization, of getting along "without the help of the corporations" is proceeding. Israel has not been able to stop it, and Kissinger's efforts to organize the industrialized capitalist countries into an economic and military force of confrontation is not going to stop it. It is this historic process that more than anything else worries the oil corporations.

The basic cause for the rise in the cost of oil products is not the prices paid to the oil producing countries. The main ingredient in high prices is the corporate profits and the city, state, and federal taxes. The rapacious Seven Thieves add a hefty "tax" at the oil well; they add more profits as the oil passes through the pipelines; and some more as it is transported by ships they own; and still more when the oil leaves their refineries; and finally the profits at the corporate gas pumps. It is a closed corporate profit cycle.

The sooner the oil corporations are driven out of the Middle East the sooner there will be peace. Israel and the Arab countries will be able to work out their differences based on United Nations Resolution 242. The sooner the Seven Thieves leave the oil fields the sooner the oil producing countries will work out the price for extracting, refining, and selling their product on a basis of equality.

The sooner the oil profiteers are driven out of business, the sooner there will be gas and oil at reasonable prices.

Truth is concrete. A vivid description of the global octopus nature of the oil monopolies is provided by 1970 company data for Royal Dutch Shell. *This company is a poor second to Exxon in size and scope.* Nevertheless, its dimensions are staggering. It has 45 operating companies in the Western hemisphere, 87 in Europe, 59 in Africa, 22 in the Middle East, and 53 in the Far East and Australasia. These 226 include only operating companies. The list does not include the big holding companies that control their operations, nor financial service companies that handle investment and other activities, nor the subsidiaries of these operating companies, which would add to the list considerably. (See Appendix A—The Global Shell Game P. 203.)

Another illustration of the pervasive and chameleon like dimensions of the structure of oil vampires is seen from a survey of selected major joint ventures, a relatively new form of operation, anticompetitive by its very nature, that takes advantage, for instance, of economies of scale—pooling of resources for large projects—for the purpose of profit maximization. For instance, in Rhodesia, Shell, British Petroleum, Mobil, Texaco, and Standard Oil of California operate jointly as Central African Refineries (PVT) Ltd. BP and Hess have a joint company in Kenya, another in Nigeria, a third in South Africa, and a fourth in the Sudan. Similar joint operations involving most of the oil monopolies of the world can be found in every continent, perhaps in every non-socialist nation. (A listing may be found in Appendix B.* P. 207)

7 • The "Patriotism" of Profits

On March 4, 1974, the *New York Times*, a paper that publishes page after page about "dissidents" around the world, but itself has an ironclad ban against publishing any "dissident" news or views against capitalism, carried a half-page article by the then chairman of the General Motors Corporation, Richard C. Gerstenberg. The article of course is a calculated falsehood from its premise to its conclusions. It is a brazen fraud. It is an article about an Alice in Wonderland capitalism.

The only excuse for the following long quotation, is that it is unique in that while it is totally false it is an open attempt to defend corporate profits and corporate exploitation, and the name of the chairman of the world's largest monopoly corporation is penned to it.

One of the legacies of Hitlerism is the use of the big lie in public relations propaganda. This statement by the chairman of General Motors is a continuation of the same technique. As you will see later this similarity between GM and Hitler fascism is not an accident of history. Because it is such a classical example of the big lie of big business propaganda we must dissect it:

* Based on Stanley H. Ruttenberg and Associates, *The American Oil Industry: A Failure of Anti-Trust Policy* (New York: Marine Engineers' Beneficial Association, 1973), "Appendix I, Selected Joint Ventures in the Oil Industry, by Regions of World, March 1973." North America not included.

As a nation we have launched ourselves on a most ambitious social agenda. We want to achieve even higher standards of education, health, and well-being for all our people. We want to abolish poverty. We want to rebuild our cities . . .

We want to give every American—of whatever color, religion, or background—an equal opportunity to become all he is capable of becoming. We aim for full employment, and even more—the full opportunity for everyone to participate in all that America has to offer.

These tasks are inherent and they are right—right for our country and for our time. But they are formidable, and sustained achievement will require a full committal of our moral and mental resources. Right now the hard fact is that the material resources essential to this task will simply not be available unless our economy stays healthy, and this is possible only if business remains profitable . . . Profits from which *come all wages, taxes, and dividends,* fuel the growth of our nation, and our future depends on the profitability of free enterprise.

Yet we are daily confronted with evidence that not enough Americans understand this. To them, the word "profit" has a grubby, selfish sound. The Vice President, Gerald Ford, noted recently, that many Americans consider a legitimate profit as a "ripoff, something that the bad guys steal from the good guys."

This lack of public understanding seriously threatens the continuation of our competitive private enterprise system.

According to a recent survey, only 3% of the American people think business as a whole is not making enough profit.

For the first time since World War II, a majority of Americans no longer thinks that companies should be allowed to make all the profits they can.

The public is wrong. Dangerously wrong . . .

We in General Motors know *there is no conflict between corporate profits and social progress.*

In addition to our efforts at General Motors to earn a profit, and largely because of our success in those efforts, *we are helping to create a better balanced system of transportation in this country and throughout the world . . .*

Profits are not applauded; they are scorned.

This is a road down which we dare not travel. . . . It is a road which leads to the dead end of excessive and irrational Government regulation—and ultimately to the end of private enterprise.

. . . we must convince people, more people than we have already, that a corporation—or any business for that matter—must *first do well* before it can do good. Better understanding of our institutions won't just happen. We must make it a goal, make it 'happen.' (Emphasis—GH).

"*Must first do well.*"It is the modern version of the old promise of "pie in the sky." Evidently the two-three billion dollar profits a year does not meet the "first do well" standards of G.M., because there surely are no signs of them "doing good."

"*We want to achieve even higher standards of well being for all our people.*"If the "We" is General Motors and "our people" are the workers, that is a cynical falsehood. General Motors, Exxon nor any other corporation has ever given one penny in wage increases unless the workers either went on strike or threatened to strike. Before trade unions workers worked 12-15 hours per day for twenty-five cents per hour or less. The human element of "higher standards of well being" has never entered into corporate operations. Well being of people and corporate profits are opposites.

"*We want to give every American of whatever color, religion or background an equal opportunity.*" The very essence of the system of exploitation is a lack of equal opportunity. The history of General Motors is a history of racism, of unequal opportunity. In these days, because of the energy crisis, we are seeing the product of unequal opportunity. Because of the history of racist discrimination against Black, Puerto Rican and Chicano workers they are at the end of the seniority lists. They were the last to be hired and are now the first to be fired. The "equal opportunity" rhetoric does not cover the Black workers in the auto foundries in the dirtiest and unhealthiest jobs in GM plants.

"*We aim for full employment.*" I am sure this makes angry reading for the hundreds of thousands of auto workers who are picking up their last paychecks, laid off because of the energy crisis, because of automation, because of speedup and a higher rate of exploitation.

But the granddaddy of all lies is the falsehood that in fact lays the foundation for all the other lies in big business propaganda and is stated in the lines "*profits from which come all wages, taxes and dividends. . . .*" There is not a speck of truth in that statement. Corporate propaganda has repeated that myth for over one hundred years. Like Hitler they hope that if you repeat it often enough someone will believe it.

They would like the public to believe that the corporations "care." And because they "care" they are in the business of benevolently passing out "wages to workers and taxes to the government." But the only thing they care about is profits. As long as no one asks where they get the profits from, they get away with this big lie.

For every nine hours on the line by GM workers, the first three are for wages. The remaining six, or twice as much, are for the boss. These six hours are the only source of the $3 billion of GM's profits. They are the only source of the $420 an hour "salary" and other booty that went to GM Chairman Gerstenberg in 1972—some $875,000.

The main purpose of the big lie is to cover up this continuous corporate rip-off. It is history's biggest rip-off and it is history's most massive cover-up.

The American people are right when they say that the GM "bad guys" in fact "steal" $3 billion per year from GM's "good guys," the workers, in an enormous corporate rip-off.

The GM propaganda department hangs on to the old fraud. Mr. Wilson, the past chairman of GM said, "What is good for General Motors is good for America." The present chairman says, "We, in General Motors, know there is no conflict between corporate profits and social progress." Nothing could be further from the truth. Because the fact is that the drive to "maximize profits" is more and more the formidable obstacle to social progress. If the $3 billion dollars went into the pockets of the auto workers they would rent and buy better homes, they would use it to give their children a better education, they would use it for better health care, for better clothing. This would be in the interest of social progress. The $3 billion that goes into corporate profits is responsible for the slums, the unemployment, the lack of hospitals and schools. GM makes $3 billion in profits and keeps most of it, because like Nixon they do not pay taxes that are in any way commensurate with their profits.

It is the quest for corporate profits, for oil profits, that involved our nation in the Vietnam nightmare, and in the Middle East monstrosity. The list of corporate crimes is endless.

One can chronicle the ills of the United States and in each instance the root cause can be traced to the mad capitalist corporate drive for profits.

The GM chairman mentions a few of these eyesores: our disgraceful transportation system, air and water pollution, and job discrimination against Afro-Americans. In each of these instances he claims that General Motors is alleviating these ills!

But as with the Hitlerite Big Lie the truth is the opposite.

General Motors has a thirty-year history of a criminal assault in the destruction of a pollution-free energy-saving mass transit system. GM buses, cars, and auto plants are the major cause of air pollution.

The waste disposal systems of this monopoly pollute the waters. Through GM's patent rights and other means, the company has thwarted the development of electric transit systems. The Washington lobby of the monopoly corporations has championed the cause of highway transportation at the expense of meaningful public transit. By maintaining a coldly calculated system of lower labor costs GM and other octopi reap multibillions of dollars in superprofits from job discrimination against Afro-Americans, Chicanos, Asians, and Puerto Rican Americans.

In discussing "law-guided processes" it is instructive to take a closer look at how these laws unfold in real life. General Motors and Exxon are the two largest industrial monopoly corporations in the world. They are good specimens for a study of the inner laws that motivate them, their moral priorities and values. They are an ugly lot.

The Ugly Americans

Fascism is the open dictatorial rule of capitalism. Thus, with Hitler Nazism, German monopoly capital clamped a brutal, murderous, fascist rule over the German people. Using Germany and Italy as a base they launched a war to establish a worldwide fascist military dictatorship dominated by German monopoly capital. If successful it would have thrown civilization back a thousand years. The drive had many small victories because monopoly capital in the United States, Great Britain, and France were united with their German counterparts in their desire to destroy socialism. Only the uncompromising greed of German monopoly capital split their ranks. The fascist drive turned into a challenge. In the United States workers were asked to make sacrifices. Calls went out to buy war bonds. Food and gasoline were rationed.

Some small-time fascist propagandists were arrested. But the real fifth column was never touched—the Axis' support in the United States that included echelons of finance capital—the biggest, General Motors and Exxon.

In goose-step fashion General Motors marched in the front ranks, providing the economic base for Hitler fascism. This was open treason.

GM and Ford also marched in the van of the big Nazi fifth column movement that provided a strong base for fascist support here in the United States.*

These facts have been revealed in a recent government-supported study by Bradford C. Snell, that at first was made available to the general public and then squelched with great haste.**

Profits is a dirty word. Blood profits is even dirtier. In the Second World War some 50 million people were murdered, while GM, Exxon, Ford, and the Rockefellers made many times $50 million from both sides.

As early as 1929 General Motors gobbled up Germany's largest automobile company, Adam Opel, A.G., and together with Ford, rapidly became an integral part of the Nazi war machine. Aided by its mass production facilities, GM's plants in Germany quickly built thousands of bomber and jet fighter propulsion systems for the Lufwaffe (Nazi air force). And to be sure of getting the biggest murderous bang possible for its buck, GM at the same time produced aircraft engines for the U.S. Army Air Corps at its plant in the United States.

* See George Seldes, *Facts and Fascism* (New York: In Fact Inc., 1943), pp. 80, 137; hereinafter cited as Seldes.

** Bradford C. Snell, *American Ground Support, a Proposal for Restructuring the Automobile, Truck, Bus, and Rail Industries*, presented to the Subcommittee on Antitrust and Monopoly of the Comm. on the Judiciary, U.S. Senate, 2-26-74 (financing provided by the Stern Fund of NY; hereinafter cited as Snell.)

As owner of Germany's largest automobile factory, GM* was a more important factor in the Axis war effort than Ford whose investments were less, though still quite substantial. GM's participation in the fascist Blitzkrieg began as early as 1935 when its subsidiaries cooperated with the Reich in establishing a new heavy-truck facility at Brandenburg, which military officers said would be less vulnerable to "enemy" air attack.**

During the following years, GM provided the Wehrmacht (Nazi army) with Blitz trucks from the Brandenburg truck facility. For these and other contributions to fascism—doing what is "right for our country and our time,"*** GM's chief executive for overseas operations in 1938 was awarded the Order of the German Eagle (first class) by the Fuehrer himself, Chancellor Adolf Hitler—an award for treason.****

Of course it must be mentioned as an aside that another member of the Big Three was also an active contributor to Nazism. In 1938 for instance Ford opened a truck assembly plant in Berlin whose "real purpose" according to U.S. army intelligence was to produce "troop transport-type vehicles for the Wehrmacht and in that year the company's chief executive likewise received the Nazi Eagle (first class).*****

In fact Ford and Hitler were kissing cousins. It is

U.S. Strategic Bombing Survey, Munitions Division, Motor Vehicles and Tanks Plant Report: Adam Opel–Russelsheim, Germany 1-26 and exhibits D-1, E-1 through E-4 (18 August 1945: Confidential: Declassified, 6-22-73): hereinafter cited as *Strategic Bombing I*; cited in Snell, p. 17.

** *U.S Strategic Bombing Survey, Munitions Division, German Motor Vehicles Industry Report 6* (11-3-45); hereinafter cited as *Strategic Bombing II*; cited in Snell, p. 17.

*** See letter of GM chairman Gerstenberg, p. 84.

**** Snell, p. 17.

***** *U.S. Strategic Bombing Survey, Ford Motor Company A.G.–Cologne 4-5 (October 1944)*; M. Wilkins and F. E. Hill, *American Business Abroad: Ford on Six Continents* 1964, p. 382; *PM New York Daily*, 9 August 1940, p. 9; cited in Snell, p. 17.

difficult to say who admired the other more. A *New York Times* correspondent noted:

> The wall beside his desk in Hitler's private office is decorated with a large picture of Henry Ford. In the antechamber there is a large table covered with books nearly all of which are a translation of an (anti-semitic) book written and published by Henry Ford.*

The Nazi treasury was heavily financed from Ford's millions, and the fiendish brutality unleashed by Hitler has been studiously copied by Ford on the production line. Murders, tortures, and beatings of workers have been a Ford hallmark to this day.**

With the outbreak of war in September 1939, GM and Ford fully converted their Axis plants to the production of military aircraft and trucks. For instance, during the first quarter of 1939, GM had already shifted its 432-acre complex in Russelsheim to warplane production. From then through 1945 this facility alone assembled 50% of all the propulsion systems produced for the JU-88 medium range bomber, "the Luftwaffe's *most important bomber*, and remained so for the rest of the war," according to authorities Wagner and Nowarra.***

The Russelsheim facility also assembled 10% of the jet engines for the ME-262, the world's first operational jet fighter, "the most important military aircraft to come out of Germany."..**** With a top speed of 540 miles per hour, it was more than 100 miles per hour faster than the American P-51 Mustang, the swiftest piston-driven fighter available for the Allied forces.

It was not until after World War II that the United States was able to develop pure jet aircraft, so that by producing the ME-262 jet engines for the Luftwaffe,

* *New York Times*, 20 December 1922.

** Seldes, pp. 122, 136f.

*** R. Wagner and H. Nowarra, *German Combat Planes* (1971), p. 298, emphasis added; hereinafter cited as Wagner and Nowarra; cited in Snell, p. 18.

**** Ibid.

GM's Russelsheim plant made a consequential contribution to the Nazi technological superiority in the air.* These were the Nazi planes that bombed and sank hundreds of ships, killing tens of thousands of American seamen.

GM and Ford subsidiaries in Germany built nearly 90% of the armored "mule" three-ton half-trucks and more than 70% of the Reich's medium and heavy-duty trucks and vehicles which, according to U.S. intelligence reports, served as "the backbone of the German Army transportation system."**

And, of course, one must not think for one moment that the Rockefeller oil kings had their regal eyes closed to the profit possibilities involved in collaboration with the Nazi monsters. The factories of Ethyl GmbH, a joint venture of I. G. Farben, General Motors, and Exxon subsidiaries, provided the mechanized German armies with synthetic tetraethyl. During 1935-36, at the urgent behest of Nazi officials who knew that Germany's scarce petroleum reserves were unequal to the wartime demands, GM and Exxon joined with German chemical interests to erect the lead-tetraethyl plants. Captured Nazi records noted that these facilities were indispensable to the German war effort: "The fact that since the beginning of the war we could produce lead-tetraethyl is entirely due to the circumstances that shortly before the Americans had presented us with the production plants complete with *experimental knowledge.*" The war document added: "Without lead-tetraethyl the present methods of warfare would be unthinkable."***

* Wagner and Nowarra, p. 370; see Jane's *All the World's Aircraft* (1945-46 ed.), pp. 204c, 205c, 271c, 272c; cited in Snell, p. 18.

** Strategic Bombing II, pp. 13-14; cited in Snell, p. 22.

*** *Hearings before the Senate Committee on Agriculture and Forestry on Industrial Alcohol and Synthetic Rubber*, 77th Cong. 2nd Sess. (1942), pt. 4, pp. 1198-99; *Hearings before a Subcommittee of the Senate Committee on Military Affairs on the Elimination of German Resources for War*, 79th Cong. 2nd Sess. (1946), pt. 10, especially pp. 1305-06; cited in Snell, p. 22.

There is no limit to corporate arrogance. There are no limits to the treachery of corporate state politicians.

After the war GM and Ford demanded, and received, reparations from the U.S. government for wartime damages sustained by their Axis facilities as a result of the Allied bombing. For example, making sure that "business remains profitable,"* by 1967 GM had received more than $33 million in reparations and federal tax benefits for damages to its warplane and vehicle properties in the former Axis territories. Ford received about $1 million mainly for damages sustained by its military truck complex in Cologne.**

Since World War II, the rebuilt Russelsheim and Cologne plants have enabled GM and Ford to recapture more than two-thirds of the West German motor vehicle market.***

Had the Nazis won, General Motors, Exxon, and Ford would have appeared impeccably Nazi. As Hitler lost, these companies were to reemerge impeccably American.

The butcheries of Auschwitz and Dachau, the annihilation of multimillions of men, women, and children throughout World War II—these and other unspeakable barbarities of Nazism have not been alien to GM, Ford and Exxon. They promoted this carnage and profited from it. Their profit interests were in direct contradiction to the best interests of the American people. The same is true today. The only item on the corporate list of priorities, now as then, is profits.

With the same calculated disregard for human welfare these corporate monsters have attacked and destroyed mass transit systems from New York to Los

* See letter of GM chairman Gerstenberg, p. 84.

** U.S. Foreign Claims Settlement Commission (GM and Ford Claims, 1967); cited in Snell, p. 22.

*** *Ward's 1973 Automobile Yearbook*; cited in Snell, p. 22.

Angeles, perfect examples of the workings of monopoly capital.

Destroyers at Work, from Dachau to Detroit

The energy crisis has brought into sharp focus the criminal nature of the corporate destruction of the best possible mass transit systems. Snell's study has documented how General Motors in particular, and the Big Three in general, with Exxon Oil have willfully and criminally destroyed more than one hundred electric surface rail systems, in forty-five cities throughout the United States including New York, Philadelphia, Baltimore, St. Louis, New Haven, Oakland, Salt Lake City, and Los Angeles. Thus does GM help "to create a better national system of transportation in this country."

Dachau to Detroit—the same process of annihilation; in wartime of people, in peacetime of their necessities of life. And of course there has always been only one reason for this devastation—the monstrous, capitalist corporation drive for profits. For General Motors the main profits are derived from the exploitation of labor involved in the production of automobiles.

One streetcar, subway, or rail transit vehicle can supplant 50 cars; therefore, eliminate the streetcars, subways, and rail transit vehicles. One train can displace 1,000 cars or a fleet of 150 cargo-laden trucks; therefore, eliminate the railroads. This is the "what is good for General Motors is good for America" logic in operation throughout the United States.

The deliberate destruction of the energy-saving, pollution-free, mass transit is not a conspiracy. It is capitalism developed to its highest stage. The destruction is the by-product of a system motivated by maximizing profits.

Here is what Snell notes about Los Angeles:

Thirty-five years ago Los Angeles was a beautiful city of lush palm trees, fragrant orange groves and ocean-clean air. It was served then by the world's largest electric railway network. In the late 1930s General Motors and allied highway interests acquired the local transit companies, scrapped their pollution-free electric trains, tore down their power transmission lines, ripped up their tracks, and placed GM buses on already congested Los Angeles streets. The noisy, foul-smelling buses turned earlier patrons of the high-speed rail system away from *public* transit, and in effect, sold millions of private automobiles. Largely as a result, this city is today an ecological wasteland: the palm trees are dying of petrochemical smog; the orange groves have been paved over by 300 miles of freeways; the air is a septic tank into which 4 million cars, half of them built by General Motors, pump 13,000 tons of pollutants daily. Furthermore, a shortage of motor vehicle fuel and an absence of adequate public transport now threatens to disrupt the entire auto-dependent region.*

What GM has done to Los Angeles it has done to all our major cities, which have become mostly parking lots, highways, and gas stations for General Motors and the Exxon oil thieves—60% to 65% of these cities' land areas is devoted to highways, parking facilities, and other auto-and-truck-related uses. And in downtown Los Angeles the figure is about 85%.**

Cars do not move in parking lots and they do not really move in the typical traffic jams and crawlways that spread across the nation. Studies indicate that city traffic moved more quickly in 1890.***

That traffic moved faster in 1890 is a good measure of the negative impact of monopoly on American life. That is about when U.S. imperialism, an outgrowth of monopoly development, first started its multinational mutilation. And today 20% of our urban population

* Snell, p. 3 (Emphasis added.)

** Ibid, p. 48.

*** W. Owen, *The Metropolitan Transportation Problem*, 1966, p. 89; cited in Snell, p. 48.

(aged, youth, disabled, and poor) lack access to automobiles. With no adequate public transportation they are locked out of jobs, and education, and other necessities.

But this is not the only yardstick. For example, there is the question of auto pollution. Forty-two billion gallons of petroleum (over one-fifth of total annual U.S. energy consumption) is consumed by motor vehicles within the densely populated 2% of the geographical area of the United States that is classified as urban. This enormous consumption spews out an excess of 60 million tons of toxic pollutants, which in turn cost urban dwellers, especially workers, more than $4 billion in economic damages.*

General Motors, of course, is well aware of this fact. On January 26, 1954, for example, E. P. Crenshaw, GM bus general sales manager, sent the following memorandum to F. J. Limback, another GM executive:

> There has developed in a number of cities "smog" conditions which has resulted in Anti-Air Pollution committees, who immediately take issue with bus and truck operations, and especially Diesel engine exhaust. In many cases, efforts are being made to stop further substitution of Diesel buses for electric-driven vehicles. . . .**

Three months later, in April 1954, the American Conference of Governmental Industrial Hygienists set a limit of five parts per million for maximum human exposure to nitrogen oxides. But Diesel buses, according to another report by none other than two GM engineers, emitted oxides of nitrogen concentrations over 200 times the recommended exposure limit. Like

* Rice, "System Energy and Future Transportation," *Technological Review*, January 1972, pp. 31-32; Samson (EPA), "The Automobile as a Social Machine," address delivered at Detroit, January 10, 1970; cited in Snell, p. 48.

** GM Interoffice Memorandum, 26 January, 1954.

a steamroller the dieselization program continued. Another GM report noted:

> The elimination of street-cars and trolley-buses and their replacement by our large GM 51-passenger Diesel Hydraulic coaches continues steadily . . . in Denver, Omaha, Kansas City, San Francisco, Los Angeles, New Orleans, Honolulu, Baltimore, Milwaukee, Akron, Youngstown, etc.*

And then there is the question of the quality of the automobile itself. Here is what the Snell study has to say about this:

> The Big Three's suppression of technological competition has precluded cost savings advances in several areas of public concern including energy, pollution control, automobile durability, and safety. It has been urged, for instance, that energy-conserving, low-emission electric and steam vehicles would help resolve this nation's acute petroleum shortage and help reduce the $6.6 billion in damages annually attributable to motor vehicle pollution. In addition, there is evidence that electric and steam-powered cars can now be produced which would cost half as much to own and even less to operate as conventional gasoline automobiles. The application of known metallurgical process could permit doubling the life of an automobile for an additional cost of (only) $36 per year, resulting in an annual savings to consumers of more than $2 billion. Effective crash-absorption bumpers, roll bars, perimeter fender protection, and other safety features have been developed which would substantially reduce both highway fatalities and the estimated $16.9 billion in economic losses ($8 billion alone in damages to vehicles) annually suffered by victims of motor vehicle accidents. Introduction of these innovations, however, would render obsolete much of the Big Three's multibillion-dollar investments in conventional body and complex internal-combustion engine production. They would also lower production costs and increase durability, thereby reducing demand, prices, and shared monopoly profits on the sale of new cars and replacement parts. There

* GM Interoffice Memorandum from F. J. Limback to E. P. Crenshaw, dated January 1954, and entitled "Operating Report—1953"; see Snell, p. 36.

are grounds reasonably to suggest, therefore, that the Big Three may have repressed these and other cost-saving innovations.*

Safety belts, crash absorption bumpers, and collapsible steering columns were already standard equipment on foreign cars when the federal government finally began prodding the Big Three to install them. Electric and steam cars were first produced by small American companies in the early 1900s. The low-emission stratified-charge engine was first developed in the late 1920s. The compact and reliable Wankel rotary engine has been in commercial production in Germany and Japan since the early 1960s. Of course, to produce these cars the Big Three would have to scrap billions of dollars' worth of technologically obsolete equipment.**

In April 1949, a Chicago federal jury convicted GM of having criminally conspired with Standard Oil of California, Firestone Tire, and others to replace electric transportation with gas- or diesel-powered buses and to monopolize the sale of buses and related pro-

* See e.g. statement of Orr in *Joint Hearings before the Senate Committee on Commerce and the Subcommittee on Air and Water Pollution of the Senate Committee on Public Works on the Automobile Steam Engine and Other External Combustion Engines*, 90th Cong., 2nd Sess., 1968, p. 63; Federal Power Commission, "Development of Electrically Powered Vehicles," in *Joint Hearings before the Senate Committee on Commerce and the Subcommittee on Air and Water Pollution of the Senate Committee on Public Works, Electric Vehicles and Other Alternatives to the Internal Combustion Engine*, 90th Cong., 1st Sess., 1967, p. 29. "A low emission Freon-Ranking engine can also be mass produced for from $20 to $25. Its relative simplicity (only two moving parts) would result in low maintenance costs." Minto, in *Hearing before the Panel on Environmental Science and Technology of the Subcommittee on Air and Water Pollution of the Senate Committee on Public Works*. 92 Cong., 2nd Sess., 1972, pp. 64, 72; Testimony of Lutz in *Hearings on the Role of Giant Corporations before the Subcommittee on Monopoly of the Senate Select Committee on Small Business*, 91st Cong., 1st Sess., 1969, p. 406; cited in Snell, p. 14.
** Snell, p. 12.

ducts to local transportation companies throughout the nation. The court imposed the piddling fine of $5,000 on GM. But this was nothing compared with the "hardship" imposed on the treasurer of General Motors, H. C. Grossman, who played a major role in the motorization campaigns and was especially instrumental in dismantling the $100-million Pacific Electric system.

*The court fined Grossman $1.00!**

Whether wartime reparations or peacetime payoffs, the government has made sure that the GM-Exxon, Order-of-the-German-Eagle entourage continues its free ride on the gravy train, yet another illustration of the workings of state-monopoly capitalism.

Here is an additional example of the government payoff to GM at the expense of U.S. workers in particular. The National Highway Users Conference, organized by General Motors as early as 1932, has compiled quite a record. Its effect has been to direct *public* funds away from rail and subway construction and into highway building. At the state level, its 2,800 lobbying groups have been instrumental in persuading forty-four of the nation's fifty legislatures to adopt and preserve measures that dedicated state and local gasoline tax revenues solely to highway building, creating a self-perpetuating fund which was legally unavailable for any other purpose. By promoting these highway "trust funds," it has discouraged governors and mayors from attempting to use these tax monies to construct anything other than highways for urban transportation. Subways and rail transit bills have had to compete with hospitals, schools, and other governmental areas for finances.

Highways and not subways have been built. From 1945 through 1970, states and localities spent more

* *United States v. National City Lines, Inc.*, 186 F. 2d, 562, 7th Circuit, 1951; certiorari denied, 341 U.S. 916, 1951; cited in Snell, p. 32.

than $156 billion constructing hundreds of thousands of miles of roads. During that same period, only sixteen miles of subway were built in the entire country. From 1956 through 1970, the federal government spent approximately $70 billion for highways, and only $795 million, or 1%, for rail transit.

General Motors, the Big Three, and the oil kings are monsters of monopoly. GM accounts for 50% of the industry's sales; 97% of domestic production is centered in the Big Three—GM, Ford, and Chrysler. Four monopolies, these three plus American Motors, manufacture all passenger cars produced and sold domestically.

The Big Three are a multinational monstrosity. They operate more than 200 automobile facilities in forty-four foreign countries and account for half of the world's total production of cars, already dominating the Canadian, West German, British, and Australian industries. Nearly one-half of Ford's workers are overseas; one-third of GM's are also abroad.

In 1972 the nine million U.S. buyers of automobiles were overcharged $1,855.10 per car in monopoly charges, overcharges on replacement parts manufactured by the Big Three's facilities and franchise dealers, and in damages attributable to motor vehicle pollution, for a total of $16.7 billion. After all, as "We in General Motors know there is no conflict between corporate profits and social progress."*

No "Conspiracy," Just Capitalism at Work

The Snell study, though it has catalogued the high war and other crimes of Exxon and General Motors in particular, believes it to be "unlikely" that there was a conscious conspiracy to destroy our nation's mass

* See letter of GM chairman Gerstenberg p. 84.

transit, but that the result was the same whether or not there had been such a dastardly scheme.*

He should have added that this makes the condemnation of capitalism all the more convincing because capitalism at its very pinnacle of development whether or not it acts as a conspiracy, nevertheless destroys all that is decent. The laws of motion of capitalist development propel it in this direction. It is an historically inevitable, law-governed process. That the inner-workings of capitalism are now exposed as a conspiracy further illuminates the dead-end direction of our social system, and points to the need to replace it with a planned economy owned and controlled by the working people of the nation—in a word, socialism.

In his March 1974 article in the *New York Times* the then GM chairman condems the American people for being critical of corporate profits. "Dismayed," he cites the fact that 35% of the American people believe that the corporations are making too much profits compared with only 3% who think it is not enough. Involved here is an elitist contempt, typical of monarchs, for what people believe in.

But the masses of the American people are right.

The Big Lie corporate statements are designed to mislead the people. One corporate executive when "in his cups" confessed the truth: "When a business spokesman makes public speeches he has to talk in terms of social responsibility and long-term profit maximization. But the truth—the deepest secret he can never admit to anyone except the lady who shares his pillow—is that he is a short-term profit maximizer."

* Snell, p.3.

part 3

• THE VICTIMS
• THE DESTRUCTION
• THE CONFRONTATION

8 • The Crisis Rip-off

The energy crisis and inflation are the instruments of the special corporate rip-off. Shortages are used to create crises and crises are used to inflate prices. This has created a circular buzz saw that keeps cutting into the living standards of the people. It is the people who are the victims of the corporate conspiracy.

The corporate propaganda machine has taken on a full head of steam. It has a single theme. The people will have to sacrifice, the poor will have to be satisfied with being poor. This is the new propaganda: abundance is no longer an assured way of life; the future of the American people is as dependent on new ways of looking at life as it is on developing new techniques.

I. W. Abel, the steel corporations' speedup expert, who occupies the presidency of the steelworkers' union, adds to the demagogy with his "equality of sacrifices" fraud.

> . . .Must we now recast our economic philosophies into the mold of a new era of economic scarcity?

So now, with the crisis well upon us, we are forced to face facts. . . .
. . .the labor movement has insisted that the needed new programs be based firmly on the principle of "equality of sacrifice."

To ask for "equality of sacrifice" between the workers and the corporations is like asking for equality of sacrifice between a thief and his victims.

Working People and Poor, the Main Victims

The crisis affects all sections of the population. But first and foremost it hits hardest at the industrial working class and oppressed national minorities, especially Afro-Americans, Puerto Ricans, Chicanos, and Asian Americans.

More than one million workers face layoffs this year alone as a result of the slowdown of production caused by the oil and gas shortage. Huge layoffs are occurring in nearly every basic industry. Current projections forecast a stiff hike in the official unemployment rate from 4.9% in 1973 to 6% by the end of this year—an increase of over 22%!

Hardest hit are some 570,000 workers in the steel industry which consumes twice as much fuel as any other. In basic steel, plans are now under way for a cutback in fuel oil supplies to the 1972 levels and for further mechanization, at a cost of some 60,000 jobs in 1974.

Current estimates predict that this year thousands of workers will be unemployed in the petrochemical industry—particularly those in plastics. The rubber industry which requires 150,000 barrels of crude oil a day just for tire manufacture is in the midst of layoffs.

And where layoffs have not been the main result, the erosion of working conditions, wages, and fringe benefits has been commonplace. For example, the International Brotherhood of Teamsters notes that the fifty mile per hour speed limit on the highways has resulted in a 20% cut in the wages of the over-the-road drivers whose pay is based on mileage covered. Teachers are being forced to work beyond their normal months as provided in their wage contracts because of the school closings due to lack of heat.

West Virginia coal miners conducted a solid several weeks-long strike against the shortage of gasoline and idiotic regulations imposed by the Governor which made it almost impossible for many miners to get to work. Both the teamsters' and the miners' struggles won substantial concessions from the government and the oil monopolies. They "found" the necessary fuel to ease the situation. Demands are rising for Nixon's impeachment on the grounds of his connivance with the oil thieves, for a march on Washington, a nationwide general strike, and for Congress to order new elections.*

For many of the nation's elderly, and especially Afro-Americans, Puerto Ricans, Chicanos, and Asian Americans, the energy crisis is a matter of life or death. People on welfare, social security, low pensions, and small fixed salaries simply do not have enough money to pay for both adequate food and home heat.

Many have been burned to death by fires as a result of makeshift attempts to fight the winter chill. For example, five children and a young Marine who tried to rescue them died in a fire that engulfed a Keystone, West Virginia home, ignited by an overheated kitchen stove.**

A ten-cent-per-gallon increase in the cost of fuel oil in Connecticut means that over 48,000 of that state's poor will be unable to afford fuel. And the Connec-

* *Daily World*, 18 January 1974.

** *Daily World*, 23 January 1974.

ticut example can be multiplied throughout the fifty states where substandard housing, lacking central heating, insulation, and conventional weatherproofing, is the plight of millions of these oppressed nationalities.

In Maine where the temperature falls to forty degrees below zero, hundreds of people were refused delivery of home heating oil because they could not pay their staggering fuel bills. Last winter fuel oil in Maine went up from seventeen cents to more than forty cents per gallon.

The independent owners or operators of gas stations have never made a living pumping gas. The stations have never even paid for the rent or cost of the lease. The corporate policy of renting or leasing stations is but another of their crimes. They are vicious revolving doors. People put their small life savings into opening a gas station, hoping to make a living on automobile repairs, only to be forced to close out in a short period. Many stations change hands a number of times each year.

Now, because of the energy crisis, the profit margins at the pump are up and the corporations are taking them over, driving the small owners out of the business.

The energy rip-off is not an isolated event. It is characteristic of the new stage of the deepening crisis of U.S. capitalism. Increasingly there will be shortages in other lines as well, not only those most related to oil and gas such as petrochemicals, or in businesses that are large users of fuel such as paper mills, but in a wide variety of basic industries.

9 • The Working Class Gets the Boot

The energy crisis is a class weapon of monopolies. It isn't directed simply against the people in general; it is directed especially against the working class. And in that sense it serves not only the oil trust, but monopoly capital as a whole.

The energy crisis was spawned in the third year of the Nixon/Congress wage-price regulation. For three years this most one-sided regulation has been used to suppress wages while permitting, even stimulating, increases in consumer prices. The wage-price regulation has been spectacularly successful for its sponsors, achieving the most rapid shift of real income from labor to big capital in the history of the United States.

The year 1973 saw the largest dollar increase in corporate profits in history, and the largest percentage increase in an advanced period of an economic upturn. And it saw the sharpest drop in real wages since the Great Crisis of the 1930s. In the first months of 1974, real take-home pay of workers declined 4% from the same period of 1973, according to the official figures of the Bureau of Labor Statistics, which understate the real decline.

How long would the workers stand for this? Throughout 1973 spokesmen of business marveled at the "patience" of workers who did not strike but accepted wage settlements of 5.5% while living costs were going up at double that rate.

This patience however, was imposed on the workers by trade union leaders who denounced the inequities of the wage-price controls, but continued to participate in enforcing them and in accepting their limitations on wages.

Something was needed to keep up the pressure, to prevent a breakout by the working class in 1974. And to that end the energy crisis is being used to create an atmosphere of emergency, of national "sacrifice." The real intent is to mobilize the synthetic public opinion of the monopoly press and TV to condemn any actions by workers that "worsen the energy crisis." For example, coal miners, facing 1974 contract negotiations, are called upon to vastly increase their productivity even though they are already producing far more coal than formerly. Decimated by speedup, by strip mining, by accident and disease, they are asked to submit to more of the same, to die young from black lung and silicosis as safety provisions are sacrificed to the energy crisis. They have been victimized by many years of rotten deals by the corrupt, murderous Boyle leadership of their union. Now they have a new and more democratic leadership and are in a better position to fight back against the fuel, metal, power, and transport monopolies that control the coal fields.

The same assault awaits the steel workers, whose rank and file are striving to overcome the crippling effects of the Abel "no strike" sellout which pledges that the union will not strike even if there is no agreement. They are told they "must not" interfere with production "needed" to provide the capital goods for "fighting the energy crisis." Abel uses his position as president to speed up production, to create even bigger profits for the steel corporations. He is by far the corporation's best little helper. In the past the trap for the drive for increased productivity was baited with the "dangers-from-foreign-competition" hoax. Now Abel is on the energy crisis kick.

During most of 1973 the big robbery of the workers' pocketbooks was carried out by the food monopolies. Consumers' movements were developing to resist, to struggle for rollbacks. And now the focus is shifted to the new monstrous robbery by the oil monopolies, linked to the electric power and gas and auto monopolies. Food prices continue to rise rapidly, but the impact is obscured by the more dramatic weekly jumps in fuel prices and the unprecedented boosts in electric power rates.

The matter was put well by S. David Freeman, a former White House energy adviser and director of the Ford Foundation Energy Project: "The energy crisis could well serve as a smoke screen for a massive exercise in picking the pockets of the American consumer to the tune of billions of dollars a year."*

Large and Rapid Layoffs

Another very serious blow against workers is in rising unemployment. The decline in unemployment associated with the economic recovery and boom of 1972-73 never got very far. The official unemployment rate didn't drop to the official goal of 4% (which corresponds to *an actual rate* of at least 8%.) And now the energy crisis is being used to cover a rapid large-scale increase in unemployment. The outgoing chief of the President's Council of Economic Advisers, Herbert Stein, who has never been known to err on the side of pessimism, predicts that the official unemployment rate will go up 4.5% to 6% in 1974. Private business and academic economists predict a rate of 7% to 8%, which would be the highest of the postwar period. Numerical predictions of job losses range between a million and a million and a half.

Auto companies have annouced layoffs totaling well over 100,000; and a total decline of 200,000 in automotive employment is to be feared. Moreover

* Quoted in *The Corporate Examiner*, December 1973.

many of these are indefinite layoffs of people whose hopes are dim for *ever* getting their jobs back. Altogether, blue collar employment in the auto industry may decline as much as one quarter this year.

The situation is equally bad in the airline industry, where announced layoffs have already affected tens of thousands, and where the prospect is for the loss of jobs by some 50,000 of the 330,000 workers. Nobody pays any attention to the gas station workers, but it's not unlikely that employment will decline another 100,000 in this industry. A rapid decline is under way in the housing construction industry whose market is reported to be "collapsing."

The impact has also been catastrophic on workers in many service industries, such as resorts, motels, restaurants, and ski lodges; all involving automotive transport of customers. These are workers with the least protection of all in unemployment insurance, union contract provisions, and access to alternative jobs.

Why so many laid off, and why so rapidly? Why was the so-called recovery of 1972-73 so shortlived? First it must be realized that even the temporary boom did not benefit the workers. It was totally one-sided and did not slow down the decline in workers' real wages, nor keep their purchasing power from plummeting rapidly in the face of the rising output of goods. The housing market was already in decline. A decline in auto sales was inevitable. Airlines were straining for permission from the Civil Aeronautics Authority to cut schedules and raise fares.

The economy was headed for a recession in 1974 and the energy crisis gave monopoly employers an excuse to adjust, in the most profitable way, to this crisis of markets. The auto companies shut down numerous plants, and retooled them for production of smaller cars. When reopened, assembly lines will run at greatly accelerated speeds, to get the same number of cars out of one-fourth fewer workers. Big price increases would ensure the same profit from a small car as was formerly gotten from a large car. We are going to get small-sized cars with big-car price tags.

The airline companies were given unprecedented authority to cut schedules and eliminate service to many places, besides being permitted to raise fares on the excuse of higher price of fuel.

At Linden, New Jersey, 2,000 of the 4,500 General Motors workers were laid off "indefinitely," including all with seniority of less than five years and ten months. Of course 60% of these are Blacks, Puerto Ricans, and women, who were only recently admitted to production jobs.* Every one of the 350 women in the plant will be laid off. The victimized workers are well aware that the gasoline shortage is contrived, as phony as the 1973 meat shortage, and designed to jack up the price.

And the energy crisis was used to distract attention from moves for a positive program to deal with the very real unemployment crisis.

Far from deploring the rise in unemployment, the monopolies welcome it in these circumstances. It fits in with their objective of keeping the workers quiet another year while they raise profit and profit margins still higher through higher prices and productivity accompanied by relatively frozen wages. They hope that unemployment will cow workers into submission, into fearing to strike for higher wages and better wages, lest they lose their jobs.

Fueling Racism

The racism of the Nixon administration and U.S. monopoly capital again comes to the fore. It's not only that Black workers suffer most whenever there is a

* *New York Times* 4 January 1974.

rise in unemployment, but the General Motors plants selected for shutdowns include those with the highest percentage of Black workers. The layoffs in the airlines will also hit hardest the Black and other minority flight attendants and other employees who have only recently broken into the industry.

The fuel crisis is used to give an extra thrust to the racist sword of Nixon and Standard Oil. It's the central cities, with their large and often majority Black populations, that are most deprived of gasoline. It's in the ghetto areas that tenements were left cold for weeks on end by avaricious landlords using the "fuel shortage" as an excuse.

The National Council of Negro Women charged that poor people, Black and other minority people are victimized as the threat of the energy crisis is used to cut into "their standard of living, including unemployment, cost of living and access to essential services." The Council called for a coalition of civil rights, consumer, and women's groups to come up with a plan of action to protect "minority groups and other individual consumers."

And the majority of the House of Representatives used the energy crisis as the cover for the most brazen piece of racist legislation in recent decades: a provision in the "energy bill" that no gasoline could be used to transport pupils for purposes of school desegregation. No mention of course of the much larger amount of gasoline used to transport pupils for the purpose of enforcing de facto segregation. No mention of course of the fact that the fathers and mothers of these minority children are forced to consume billions of extra gallons of fuel because they aren't permitted to live where they work. The Senate may yet, at this writing, cancel out this provision of the House of Representatives. But no one of conscience can vote for the reelection of any member of Congress guilty of supporting this vile racist piece of legislation.

It is not unusual that Afro-Americans, Chicanos, Asian and Puerto Rican Americans have been especially victimized by the energy crisis. This is part of a historic process of racist barbarism. Racism has been a permanent feature of U.S. development.

For example, during the post-Civil War Reconstruction period a struggle developed over the title rights to the enormous plantations in the South, which included vast mineral-rich territories, especially of petroleum and natural gas in Texas, Louisiana, and Oklahoma which today account for 45% of the nation's principal mineral production. The robbery of the Reconstruction period had a special racist edge.

Involved here was an alliance between the capitalist interests of the North and the former slavocracy in the South. The businessmen were quite willing to trample on the rights of the actual tillers of the land, the actual producers of the wealth, as long as the system of wage labor, a prerequisite for capitalism, remained dominant. And of course the landlords, like the landlords of today, were quite happy with the fact that they had to do nothing but sit back and gobble up the profits.

This collaboration in criminality proved to be more powerful than the growing alliance between the ex-slaves and the growing, youthful working class.

It soon developed that the rich areas were not to be turned over to the four million freed men, women, and children, the rightful inheritors, the ones most deserving of the land. After some initial successes based on heroic struggle, the former chattel slaves were forced back upon the plantation under labor contracts or as tenants and sharecroppers. And the boss typically was the same ex-slave owner. The roots of Exxon and the Rockefeller oil empire are in slavery and the Reconstruction swindle.

New Monopoly Methods and Bigger Bonuses

The energy crisis has become the biggest smoke screen for a multibillion-dollar rip-off. Monopoly control of product cycles from raw materials to retail checkout counters makes it possible for them to manipulate supply and create shortages and crises. The crisis is then used to create an atmosphere of panic. When that happens the situation is ripe for the corporate blitzkreig.

It has become the excuse for the Department of Justice to drop most of the antitrust suits. Consumer protection is shelved, environmental protection is ditched, rent controls are set aside, and prices and taxes escalate. Anything that stands in the way of maximizing profits is dumped. All the backlog of restrictions and safeguards are scrapped in the name of the energy crisis.

Once the corporations have had their way some of the shortages will disappear—until the next time. Shortages and crisis are new monopoly weapons against the people, to be used as instruments for bigger profits.This is a new wrinkle, a new stage in the development of state monopoly interlocking capitalism.

The basic source of corporate profits is at the point of production. The employers use every possible device to tighten the screws of exploitation in the factories and in the mines. The steel workers are told they must produce more steel, work faster, because the oil industry needs more steel pipe. The oil workers are told they must speed up production because the nation needs the oil products. In modern society there is no industry unrelated to the overall economic process.

In a new way state monopoly capitalism is a factor in the unprecedented drive to increase labor productivity, and the state operates as an open partner, in fact, the driving wedge. Labor productivity is paced by the new technology which, more than replacing the human element in production, has become an instrument for dehumanizing speedup. This brutal intensification of labor continues unabated, without serious resistance or counterattack in sight, and is thus a continuing setback for labor.

The new technology is providing the monopoly corporations with an instrument of production and, even more, with a new demonic instrument of speedup. Because parts of the production line are automated—every second of the working hour is automated and geared to the maximum of human endurance. This factor has emerged as one of the new decisive unsolved problems for the working class.

Workers in mass production industries have become cogs, robots in an automated process in the most controlled, integrated, brutal system of exploitation ever devised. They are victims of a mass production system run and dominated by gauges, by control indicators, count-rate meters, events-per-unit time meters, load indicators, strain gauges, pressure indicators, vibration and stress analyzers, and a hundred more such control mechanisms. But these devices are not used to measure the stress, pressure, or load that the workers carry. They check the stress and pressure on metals, their maximum capacity and speed in relation to the rate of production. These devices are used to drive the workers to the very limits of human endurance, which show up in the shortened life spans, in the medical records, in the criminal increase in industrial accidents.

These changes in the production process are bringing with them some changes in the work force. Blacks, Chicanos, Puerto Ricans, women, and young workers, by and large, are not in the button-pushing departments of the industrial process. Increasingly they are in the departments where the automated machines and the human hands mesh into a speeded-up, automated production line. These are the departments where the intensity of labor is now the highest. They are the most highly exploited. The automated lines especially need young hands and young unused nerves.

In the automobile industry 60% of the workers are now below thirty years of age, with large sections of eighteen to twenty-year-olds. There are also thousands of women workers in the shops of GM, Ford, and Chrysler.

Racism still keeps the Black workers in the foundries and in the other lowest paid, dirtiest jobs. But more and more, especially the young, Black workers are in the high intensity, speeded-up production lines. This adds a new dimension to racism on the production line. This adds a new importance and need to give special attention to young workers, to women workers and to their special problems in industry. On top of all these problems we must also add the energy crisis layoffs previously discussed.

State monopoly capitalism is an efficient profit-squeezing machine whose policies are designed to give monopoly capital an ever larger portion of the national product. Each new stage of its development sets the objective patterns that will lead to an intensification of labor productivity and to sharper class confrontations.

Endless Inflation

The inflationary trend is a factor the working class will have to face from now on. There will be some ups and downs but the question of how to keep up with or how to close the widening economic gap will remain a critical issue. This situation is devastating to millions of people. Inflation has destroyed the prospects of economic security for the millions living on fixed incomes, which includes those on social security, welfare, pensions, people living on savings, and the millions of unorganized workers whose wages remain basically static. Inflation has a brutal, racist face. In the ghettos and barrios inflation is a deadly killer. It sentences the oppressed minorities to a constantly declining standard of living—to a bare existence standard of living.

The basic essence of inflation is that it takes from the working class and the poor and adds to the profits of the rich. What the corporations let pass at the paycheck window they take back at the checkout counter. This continuous inflationary spiral is related to the new levels of state monopoly capital and to the new role of the state and the banks.

The growing system of credits on all levels adds to the inflation. The state monopoly structure is able to resist price reductions during periods of oversupply or recession and maintain the inflationary trend. Recessions do not act as a means of correcting prices despite oversupply. Once all business transactions are made within the context of the acceptance of continuous inflation then each transaction becomes a built-in inflationary stimulant. Inflation has become a way of life under world capitalism. Because of its anarchism and decay it is not able to control this development. Inflation is a double-edged sword for world imperialism. It postpones economic crisis, but it also undermines the stability of capitalism. The system itself twirls on the inflationary spirals.

The two critical areas—inflation and the automated intensification of labor—present the working class with new and difficult challenges. They have resulted in a crisis of everyday living and a crisis of everyday work load. Because they are for the most part unchallenged, these two areas are giving the corporations skyrocketing profits.

It is also in these two areas that the monopoly corporations most jealously guard their so-called management prerogatives as some sanctified, sacred, corporate territory. Their position is that these areas are not negotiable. They are not mentioned in the labor-management contracts.

The trade union machinery is not structured to take up this challenge. Managements of most industries are ready to make concessions in most other areas as long as these two go untouched. Though missing from national contracts, they make their appearance in what are called "local issues." They are local only because they are not taken up on an industrywide basis. They are mainly related to the intensification of labor—grievances on working conditions, on job security, speedup, and crippling accidents.

With each contract the number of unsettled local grievances keeps growing. For the workers at General Electric the list has grown in two years from 34,000 to 65,000 unsettled grievances. The workers at General Motors have 67,000 unsettled grievances, Ford workers have 20,000, and the Chrysler workers have 20,000. This represents an accumulation of resentment and anger that will explode.

Automobile production is an example of how the speedup of workers takes place. For 1973 the production line was speeded up to produce an additional five cars per hour, without adding any additional workers. This added forty to fifty cars per day for each line without a penny of additional cost. This has become a trend in the mass production industries. This is where the new maxiprofits are coming from.

Over the bumper harvest of profits of 1972, U.S. Steel increased its yearly profits by 107.5% ; Republic Steel by 101.5%. These are massive profits, but Abel, the head of the Steel Workers Union, is a full-time drumbeater for more speedup and higher profits.

Thus, the new level of state monopoly capital results in a new level of exploitation, a new level of speedup, a new level of racism. It also results in a new level of class collaboration, all ending up in a new level of corporate profits.

These new developments have brought on a sense of crisis in the trade union movement. There is a feeling among workers that the trade union movement is boxed in, a victim of the policy of class collaboration.

The crisis of the trade union movement is deepened by the leadership policy of doing nothing about united action. Each strike is on its own. Meany has made his boast of never walking a picket line a policy of principle for the top leadership of the trade union movement.

The Watergate crisis reveals the corruption, the build-up of a police state structure in government. The energy crisis reveals the further development of monopoly capitalism. These two processes are not unrelated. In fact they have an interlocking relationship. They are two sides of one overall process, a further decay of capitalism.

10 • Racism and Oil Profits

The oil industry has always been the exclusive preserve of the very rich. The stolen loot, starting with the colonial acquisitions, passed on to each new generation of oil thieves. Their family roots go back to the supporters of British colonial rule.

By any yardstick of justice the riches of nature belong to the people who inhabit the land. But that is not the case in the United States. The minerals and oils beneath our feet have been stolen—confiscated by the big corporations. The thievery started with the colonial occupation. The Indian peoples were the first victims.

Our history is a history of corporate plunder. That which rightfully belongs to the people now belongs to a handful of large corporations. They not only have the mineral rights to most land areas but they have pushed through laws that make their "right" to nature's riches the "first right." When they have wanted the iron ore beneath cities and farms they have, without compensation, bulldozed them out of the way of their profits.

Racist From the Start

The corporate rape of the people's resources has always had a sharp racist edge. The monopolies have always effectively barred Blacks, Chicanos, Indian Americans, and Asian Americans from holding mineral rights. The Civil War and the Emancipation Proclamation did not open the mineral rights door to Black Americans. The federal government held title to large tracts of mineral-rich land, but it was put on the market in such large lots that only the very wealthy could bid on it. This practice continues to this very day.

The American Indians have been victims of four hundred years of the most brutal, inhuman oppression. Not the least of this oppression has been the theft of mineral- and energy-rich lands from the earliest Americans by the oil robber barons.

A recent example of this oppression has been the murderous onslaught against the Hopi Indians in northern Arizona by the Peabody Coal Company. Peabody Coal, a subsidiary of Kennecott Copper Corporation, is the nation's largest producer of coal. This industry is dominated by the oil thieves. Of the top fifteen U.S. companies that produce coal, six are oil and gas companies. These include Continental Oil, the second largest coal producer; Occidental Petroleum, third largest; Standard Oil of Ohio, tenth largest; and Gulf Oil, thirteenth largest.*

In the most barbarous fashion Peabody has been tearing apart 14,000 acres of the Hopi Black Mesa land in a pollution-laden, strip mining operation that has cast a deadly pall over the whole area.

The stench from strip mining emitted by the oil-dominated coal kings pours out from more than one out of every three tons of coal that the nation produces. In this process huge electric shovels gouge out the top soil to get at the readily accessible seams of coal. Despite state laws that supposedly require the mine operators to reclaim the land by grading and restoring topsoil, in practice, the strip miners leave a wasteland—barren, treeless, and unusable.

Thanks to Peabody Coal, more smoke has filled the area than from all stationary sources in the City of New York, according to the Coalition of the Environmental Defense Fund, National Wildlife Federation, and the Native American Rights Fund.**

Net "gain" to the Hopi Indian?—an estimated twenty jobs under brutal, inhuman conditions and for poor wages!

Before there was a United States of America in the legal sense, the treaties with the Indians were made by the colonial powers which vested authority in a governor of a colony. These governors for the most part apportioned the land to their relatives and other favorites. Even in the earliest times there was an enormous amount of fraud, corruption, theft, and out-

* *Daily World*, 15 November 1973

** *New York Times*, 15 May 1971.

right massacres of Indians—all concerned with taking away the Indians' land rights.

The Indian stage of civilization was a form of primitive communalism in which there were no property rights as in future stages of civilization. There was no private ownership of land; everyone owned it.

The Indians with their superior, more ethical concept of property, easily fell victims to the vile amorality of many of the European settlers, the bearers of capitalism, an import to these shores. The treaties the Indians made with the settlers in which they sold their land were generally out-and-out robbery, as with the oil corporations and the public today.

The history of the Indian lands is a history of broken governmental treaties; broken firstly by the colonial rulers and then especially by the federal government aiding and abetting, by force and violence, the flagrant theft of the lands by the actual settlers. This history has been a one-way process of robbery and murder of the American Indians.

After the American Revolution the land treaties with the Indians were made with the United States government. The Louisiana Purchase of 1803, Florida in 1819, Annexation of Texas in 1845, Mexican Acquisition of 1848, Oregon in 1846, Alaska in 1867, Hawaii in 1898 were federal government activities; many of which were military conquests.

When these territories applied for and were granted statehood by Congress, the territories were ceded to the states from the federal government and then fell under the jurisdiction of the individual states involved. But this was after the basic land question had already been determined between the federal government and the other colonial powers (e.g., Louisiana Purchase) and the Indians. The original land settlement by the Europeans actually occurred in most of these territories before they became states, a certain amount of economic development being a prerequisite for statehood in the first place.

The price of purchasing the land from the federal government was quite high so that generally only the wealthy with aristocratic lineage could afford to buy it. This was true from the early beginnings, e.g., the Land Acts of 1796. The highly touted Homestead Act of 1862 provided cheap public land which was by and large of poor quality—and not much of it at that—compared with the land that already had been distributed.

The original European settlers usually got their land directly from the federal government, mostly through purchase, since the land was generally settled before the territories became states. Involved here were favored hereditary connections and direct theft of the land from the Indians (i.e., open violations of the agreements between the federal government and the Indians), a very widespread general practice.

As late as 1860 the greater part of resources devoted to manufacturing was under the control of proprietorships not of corporations. This was typified by the oil companies. It was not until the period of the Civil War that an oil strike of any significance occurred with the discovery of oil by Drake in 1859 in the northwestern corner of Pennsylvania. And it was not until the 1890s that the westward movement of oil discoveries took place.

Therefore, the corporations, the oil corporations in particular, as a rule did not get involved until statehood was generally established. Much of their dealings were with landowners who acquired title rights from a particular state, which in turn, had been ceded the territory by the U.S. government.

But by 1969, 517 million barrels of oil and 2,653 billion cubic feet of natural gas were produced on government lands, amounting to about 15% of the total oil and 12% of the total natural gas produced. Payments to the government for that rich haul came to only $129 million, or the equivalent of about 20¢ per barrel of oil. The amount of oil and gas taken from government lands is increasing at the rate of 10% per year, while that taken from privately owned lands is decreasing.

However, there are enormous areas of land, as in Arizona and especially Alaska, that are still federally owned. There are 375 million acres in Alaska, but the Indians, Eskimos, and Aleuts who are among the most starved, ill-clothed, and ill-housed peoples of the nation receive only 40 million acres of the poorest lands under the terms of the land claims act signed by Nixon in December 1971. The remaining 90% of the land goes to the federal government to give away to the monopolies. In the fabulously rich North Slope oil region native Alaskans get only surface or hunting rights on most of the area. The underground oil treasure is left for the Gulf, Exxon, and other pirates.

Not one cent of the land claims settlement, which was less than $1 billion, went directly to the people who needed it—the 55,000 impoverished native Alaskans. Instead, $462.5 million is to be paid out over an eleven-year period to shareholders of twelve Alaskan corporations, who will dispense it somewhat like the government poverty programs.*

The "theory" was that the money would trickle down. But the bureaucratic courtesy of Uncle Sam has frozen the benefits. Pennies trickle down to the people, the dollars gush into corporate treasuries.

Racist Policies Abroad

The whole energy crisis reeks of racism on the international level as well. The collaboration between

* *New York Times*, 5 December 1971.

U.S. imperialism and the chief European racist colonial government of Portugal reached a new high. Portugal gladly permitted the United States to use its bases for shipment of military supplies to Israel, the aggressor state, something none of the other allies of U.S. imperialism dared to do. And the Nixon administration of course continued to supply weapons and whatever else Portugal needed for its campaign of suppression against the people of Angola and Mozambique.

And consider the collaboration between U.S. imperialism, the Gulf Oil Co. and the Portugese colonialists. Gulf is now getting large quantities of oil from Angola. It increased production in the Cabinda fields, the main fields of Angola, by 9% in the very first month of the Middle East fighting.

Here is the story, as reported by the *London Observer*:

> The Gulf Oil Company has discovered a major new oilfield off the West Coast of Africa. The field is believed to be one of the biggest finds in recent years, and Gulf officials are already referring to it privately as a "new Kuwait."
>
> The find has not been made public because of the haunting political problems. It lies just off the coast of Cabinda, a part of the Portugese colony of Angola, where guerrillas of the MPLA are operating against Portugese troops. *To get the oil out, Gulf has been collaborating with the Portugese in an attempt to keep the guerrillas out of Cabinda.* The MPLA has penetrated deep in Angola and has obtained sophisticated new weapons from the Soviet Union. In recent weeks it has been active inside the Cabinda enclave.
>
> The total size of the field is still a closely guarded secret, but the oil industry sources believe it may be very large. Some oil industry sources even suggest that President Nixon's serene attitude to the Arab boycott can be explained by the size of the Cabinda find.* (Emphasis mine—GH)

And racism is used to arouse the public against the

* Reprinted in *Cleveland Plain Dealer*, 25 November 1973.

Arab people, and to deflect popular anger away from the real culprits responsible for inflation and shortages.

In the 1972-73 crop year the Soviet Union made substantial purchases of grain in the United States. This was very beneficial to the farm areas, to the farm equipment industry, and to segments of the transportation industry. It helped get rid of burdensome grain surpluses and helped end abnormally low grain prices. But the grain and food monopolies saw an opportunity to make a killing and blame it on the Soviet Union. After the Soviet Union had stopped ordering, and the companies had bought up the grain from the farmers, a false tale of a grain shortage in the United States was whipped up. It was used as an excuse for pushing the price of grain sky-high, and then raising prices of meat, eggs, and bread—all basic foods—several times more than could possibly be justified even by sky-high grain prices. It was a $30-billion rip-off of the U.S. consumers by the American food trust. And it was all blamed on the Soviet Union for buying a billion dollars' worth of U.S. grain at regular market prices.

This canard was kept alive—and is still revived almost monthly by the *New York Times*—no longer to defend the food trust, but as part of a persistent campaign to reverse the move toward detente, to get the country back on the dead-end track of the cold war.

And now we have the same device used against the Arabs. Their very mild use of oil as an economic weapon against Israeli aggression and against the imperialist supporters of that aggression is grasped by the oil trust as a heaven-sent opportunity to invent a fuel shortage and put over the same kind of multibillion-dollar price gouge as the food trust did. This is likely to end up costing the American consumers as much as the food rip-off. And, as with food prices, there is no prospect of a later cutback as long as monopoly control of oil remains.

U.S. Imperialism Cries "Foul"

Moreover, U.S. imperialism is striving to build up resentment against the Arabs to gain support for its aggressive policies in support of the oil monopolies. Kissinger and Nixon and Schlesinger made repeated threats of counteraction against the Arabs and newspaper columnists friendly to them voice what they merely implied—the threat of direct U.S. invasion of Arab lands.

The main effort of U.S. imperialism is to rebuild the weakening NATO alliance into a new front of so-called oil consuming countries against the Arabs, a front which will dictate to the Arabs and other producing countries how much oil to produce, whom to sell it to, and what compensation to pay the cartel for nationalized properties. As Nixon said openly, they threaten to repeat the economic and political warfare that U.S. and British imperialism used against Iran twenty years ago.

Jamil N. Baroody, Saudi Arabian representative to the United Nations in a recent debate attacked U.S. officials who accused the Arabs of blackmail:

> Mr. Baroody said that he had read where Mr. Kissinger had threatened countermeasures. However, he noted the U.S. Secretary of State did not have the guts to say what they were. He recalled that, some years ago, an American friend of his, a Roosevelt (Nicholas L.) had left the government service to serve as undercover agent in an oil company, created in a country which had nationalized the oil industry. The government of that country was toppled, he noted.
>
> The CIA, with its large budget, engaged in subversion, coups d'etat and cloak-and-dagger methods, he observed.

He warned that if subversion was used against the Arab countries, "you will never get a drop of oil."*

But considering the present alignment of world forces, the attempt of Kissinger and Nixon to knock together a new imperialist alliance against the oil producing countries would objectively mean to take on virtually the whole third world and the socialist countries simultaneously.

It is doubtful whether many other capitalist countries will want to attempt an adventure with such dubious prospects, especially since they themselves are being victimized by the manipulations of the Anglo-American-Dutch oil cartel.

Imperialist countries involved in oppressing and exploiting other nations and peoples are predominantly white. Most of the oppressed peoples and nations are nonwhite. Throughout its history imperialism has maintained this structure of oppression with a special racist essence.

The struggles of the oppressed are against an imperialism that is ideologically sustained by racism. The racism U.S. imperialism injects and uses on the domestic front, is also the pattern of its world operations. This creates a unique relationship between the struggle against imperialism and the struggle against racism at home and abroad.

In some areas imperialism has been forced to give up its open colonial-military method of oppression. The neocolonialism is more sophisticated; the oppression is more indirect. Reflecting this change the racism is also more "refined." But the aim of neocolonialism is the same, and the essence and purpose of racism are the same.

For proof that imperialism uses neocolonialism when necessary as a maneuver and retreat, one has only to see the open brutality, the open racism in

* United Nations Press Release GA/EF 1516, 4 December 1973.

South Africa, in Rhodesia. Vietnam stands for all time as a symbol to the world of the murderous, bestial nature of U.S. imperialism. The Exxon and Gulf oil drills in the seas off South Vietnam remind the world that the carnage, the massacre of women and children, the napalm holocaust, the 50,000 young Americans killed, are crimes that must be placed at the doorstep of U.S. imperialism. The present neocolonialist policy in Vietnam is a continuation of its racist-imperialist aggression by new methods.

A more basic resolution to the energy crisis in the United States is related to the struggle against imperialism and racism. The oil producing countries are reacting to both: they are not going to sell their oil at colonial prices, and they are not going to accept the racism of the imperialist countries.

11 • Pollution and Profits

Corporate plans for the energy crisis included creating and using an energy crisis backlash. Planned shortages, monopoly capital's new weapon to create the backlash, are also a basic instrument of the Nixon administration to raise prices until the poor sections of the people cannot afford to buy. This results in a decline of demand, and thus supply is built up. It is a policy of lowering the standard of living, of using hunger to cut down demand in order to balance it with supply.

One of the areas where the whipped-up energy crisis backlash has been used most effectively is in the blitzkrieg against pollution standards and controls. Measures that provided some minimum protection to safeguard a livable environment have been ploughed under. The Nixon people pushed through the Alaska pipeline minus the basic requirements of environmental protection. Strip mining is ugly and wide open. The movement to safeguard the environment is on a treadmill and slowly slipping backwards. The dominoes of environmental protection are being knocked down one after another.

A remark by a vice president of the Consolidated Coal Company, one of the worst polluters in the country, is a reflection of the attitude of the corporations generally who insist that environment is not their responsibility and that people who fight for it should be treated as "criminals," as "eco-freaks."

> Conservationists who want strip miners to restore land are stupid idiots. They are "commies" who don't know what they are talking about. It is our duty to knock them down and subject them to the ridicule they deserve.*

Energy and a livable environment have merged into a single, critical problem. Henceforth, the extraction and use of energy without simultaneous and planned activities to guarantee a livable environment is a dead-end street.

Corporations are motivated by "maximum profits." The future of energy and a livable environment will depend on maximizing human needs. Capitalism and social progress are on a collision course.

Energy and the environment face both immediate and long-range crisis. They are interlocked. On the environmental clock it is later than most people think. The pollution goes on every second, minute, hour —every day of the year, year in, year out. The most dangerous pollutants are unseen, unnoticed. They are accumulated and they have a point of no return.

Profits and Their Bitter Fruits

In 1973 the conspiracy going on for several years between Nixon and the oil thieves began to bear its

* *Yonkers Statesman*, 7 February 1974.

bitter, poisonous fruits. The price of domestic crude oil shot up from $3.45 a barrel in 1972 to more than $6.00 average by the Spring of 1974. The wholesale price of gasoline advanced 50% during this same period; profits of the major U.S. oil kings were fabulous.

But nowhere were the poisons more apparent than in the increased pollution that Nixon and his gang deliberately fostered. And the pollution was not only verbal and ideological.

Total outlays by the Environmental Protection Agency (EPA) in fiscal 1973 were $1.1 billion, while as far back as 1968 the amount of damage caused by air pollution alone cost $16.2 billion according to another administration agency.*

Throughout 1973 the Nixon-dominated EPA granted the automobile moguls three extensions for meeting emission standards, and a year's extension, from 1976 to 1977, was set for meeting emission standards for nitrogen oxides.

In his energy message in January 1974, Nixon called for a two-year postponement of the already insufficient auto emission standards that were to have gone into effect on 1975 model cars. Originally, the Clear Air Act of 1970 required a 90% reduction in auto emissions of hydrocarbons and carbon monoxide for the 1975 model cars from the levels produced by the 1970 models. However, at present the EPA has set "interim" 1975 standards that represent a reduction of only about 65% to 70% from the 1970 levels. The Nixon syndicate wants to freeze these woefully inadequate interim standards for two more years.

The General Motors-dominated auto industry, in turn controlled by the Morgan and other groups of finance capital, is not only the nation's largest single air polluter, but it also gobbles up a major piece of the domestic energy pie.

* *Fourth Annual Report*, Council on Environmental Quality, p. 78.

A serious solution to achieving clean air requires a kayo of the corporate moguls. It requires a people's public transportation system. But this is farthest from the thoughts of the Nixon syndicate despite its recent token proposals for increased aid to public transportation—a response to the shortages of gasoline at the gas pump. For example, in fiscal 1973 it spent less than $400 million on public mass transit and less than half a billion is being spent currently, but billions of dollars go to the Pentagon budget whose "cost overruns" on major procurement programs alone have exceeded $35 billion. *Not a penny* has been spent by EPA for development of clean fuel-saving cars that are operated entirely or partially by electricity. One such model, operated both by electricity and gasoline gets 50% more miles to the gallon than the average car and can meet the strictest emission standards, claims a Dr. Wouk and his family who have personally invested $300,000 in the venture. But "with a little modification in a later version," Dr. Wouk contends, "we could run an all-electric model and solve the housewife's fuel problems. She could drive ten miles to the supermarket just on the batteries, and when she got home, plug into an outlet in the garage and recharge the batteries for the next day."*

However, capitalism has not and will not "plug into" the new technology to meet the people's needs.

Pollution's deepening stench is not only due to the larger droppings of the Nixon rat pack; it is also a manifestation of capitalism's refusal and inability to cope with its deepening crisis, with the new technology. It is the central theme in my book *Ecology: Can We Survive Under Capitalism?*,** in which I quote Frederick Engels as follows:

> As long as the capitalist mode of production continues to exist it is folly to hope for an isolated settlement of the

* *New York Times*, 27 January 1974.

** Gus Hall (New York: International Publishers, 1972), p. 90.

housing question or of any other social question affecting the lot of the workers. The solution lies in the abolition of the capitalist mode of production and in the appropriation of all the means of subsistence and instruments of labor by the working class itself.*

How Socialism Solves the Problem

Moscow is one of the world's largest industrial cities both in area and population, but its air is the purest compared with any capitalist city of similar size. There are over 7,000 purification plants at Moscow's heavy industries. The boilers and central heating power plants that once used liquid or solid fuel have now switched over to gas which is relatively pollution-free. The health of the city is vigilantly supervised by the sanitary services.

Despite the swiftly growing industry and steadily expanding use of automobiles, air pollution in Moscow has declined by over 85% since 1948. Dozens of laboratories specializing in analyzing the environment operate today in various districts in Moscow.

Although the city is surrounded by a large greenery belt—a natural filter that purifies and sanitizes the air environment—*about 30% of the city's area* is occupied by greenery plantations. This is an additional oxygen reserve of the capital. These greenery plantations include huge forest-parks. The Izmailovo Park, for instance, is almost four times bigger than Central Park in New York and eight times larger than Hyde Park in London. In all there are 100 parks, 660 landscaped thoroughfares, public and other gardens in Moscow today.

When new factories are built care is taken to ensure that they do not exert an adverse effect on the host community. Every factory is surrounded by a green

* *The Housing Question* (Moscow: Progress Publishers, 1970).

sanitary protective zone ranging in width from 45 to 900 yards.

In the USSR, under socialism, they use their profits to depollute or remove offending plants from the cities. Short-run profits are not decisive when the general public owns the entire economy. Though many could be cited, here are just two small examples.

A big oilcloth plant near Moscow was told that a new housing area was going to be developed nearby. The experts agreed there was no way to remove the bad stench from the plant processes. So the Moscow City Soviet (council) decided that the plant must be moved and assigned land for it.

At a big oil and gasoline refinery inside Moscow the stinking sulphur fumes were removed when new devices were installed by the chemists. At the same time the plant removed the lead from the gas so that none of that poison could pollute the Moscow air.

These measures cost money, but the "dividends" go toward everybody's health in the Soviet capital. What is happening in Moscow is happening in every city in all of the lands of socialism.

Under socialism there is no room for big auto monopolies such as General Motors, Ford, and Chrysler. There is a good public transportation system so that the private automobile is not the necessity it is in the "free world." The development and constant improvement of public transportation is at the top of the priorities list.

The number of automobiles, however, is on the rise, along with the improvement of the general living standards, providing an extra convenience to the average worker. Even so, as cars multiply the car pollution standards become stricter.

In the Soviet Union some 3,000 cars have changed over to compressed gas as fuel in place of gasoline. The current 1971-75 Five Year Plan calls for the conversion of nearly all of the nation's taxis, trucks, and buses from gasoline to liquefied or compressed natural gas. Some large vehicles are changing over to four-stroke motors which are less polluting. The traffic is kept moving even in rush hours preventing big pollution jams.

Long ago Soviet engineers saw that diesel buses would literally choke downtown streets. They set about solving electric bus problems. Today electric buses by the thousands can be seen moving around Moscow in the coldest winters—fast, silent, stench-free, and comfortable. Soviet experts predict that electric automobiles will soon replace the gasoline engines. At the very moment when Soviet cities were converting to electric buses, Chicago converted from electric buses to gas-burning buses.

More than 3,000 trucks burning natural gas are already in use in Moscow. The USSR has enormous reserves of natural gas. There is no problem of supply. Also there are no entrenched oil company interests to protest a switch to natural gas in the USSR's socialist economy.

The Soviet Union was the first nation to issue a total ban on the use of leaded gasoline in cities and recreation areas. Lead, added to gas, poisons the air, as those of us living in the United States know all too well. But just plain gasoline without lead when ignited inside a car engine not only burns but also detonates—causing a "knock"—and this can ruin the engine. Lead dampens the detonation process.

But there is another ingredient that can be used in place of lead as a dampener—manganese. For over a year all gas stations in the city of Orel in Central Russia have sold gas with manganese rather than lead. The results have shown that manganese is just as good a dampener as lead, if not better, and does not create the toxic exhaust emission as does lead. The USSR has a good supply and can think about replacing all lead in gasoline with manganese in the future. The United States, on the other hand, has virtually no domestic reserves of manganese—essential to the manufacture of steel—and is completely dependent on outside sources. But there is no shortage of manganese on the world market.

The big Lenin Truck Works in Moscow which makes ZIL trucks is experimenting with truck engines using precombustion-chamber flame ignition systems designed by Academician Nikolai Semyonov. Passenger cars made in Gorky, the USSR's main automaking city, now use Semyonov engines. These engines insure almost complete combustion of all fuel so that there are no waste products of incomplete combustion—a major component of U.S. urban smog.

The USSR last year built a number of experimental electric cars but found that their range before recharging their batteries was around thirty miles. Today the test cars use a combined system. They have a small internal-combustion engine used to recharge the electric batteries while the car is moving. Soviet authorities contend that the new combined system gives the test cars the same range as a conventional car. A similar system in the United States is barely off the drawing boards. The government so far has only started to test these vehicles and private industry will not develop them because of the changeover costs. *The obstacle to pollution-free automobiles is not a lack of knowhow.* The obstacle is corporate profits.

Achievement in the area of pollution control spreads over the entire Soviet countryside. Without doubt the most dramatic story is that of Lake Baikal, 400 miles long, 50 miles wide, and 5,750 feet deep. Its volume equals all five of North America's Great Lakes. Yet the waters of Baikal are so transparent that you can see to a depth of 125 feet.

My book on pollution and the environment was

written two years ago. Many of the passages are especially relevant today:

> What happened there at the earth's most ancient and deepest lake was told, not by a Soviet publication, but by Farley Mowat in the *Boston Globe*.
>
> The lake, Mowat wrote, contains almost a fifth of all the free fresh water in the world. Its depth is more than a mile, the waters "fantastically clean," and Baikal has more than a thousand species of plants and animals that are found nowhere else.
>
> Mowat continues: "In 1962, the economic planners in Moscow decided to build five gigantic cellulose and wood-chemical plants on Lake Baikal. In 1964 work on the first two began. At this juncture something truly remarkable occurred. *Pravda* and *Izvestia*, having proudly announced the birth of the gigantic new production complex at Baikal, were inundated by letters of outrage. As the two plants neared completion the intensity of the storm strengthened. An elderly, much-respected Moscow writer described what followed: 'The word Baikal became a rallying cry even to people who knew very little about it except its name. They were acute enough to see that finally the high priests of progress-through-production had to be brought to their senses. . . . For a while the authorities who had designed the cellulose combine tried to drown out the protest. . . . There were some threats. . . . The plants were completed and began operations. . . . Within three months there were reports of fish dying in Baikal and even of people getting sick from eating fish caught in the Angara. The fight of the people to save the lake became more furious and then, quite suddenly, the authorities gave in. The plants were closed.
>
> "To the Western mind the scope of the victory seemed staggering. In 1967 the Presidium of the Supreme Soviet of the USSR voted to make the entire Baikal region—the lake and thousands of square miles of surrounding territory—into a national park. Extensive reclamation projects are under way to restore tributary streams and riverbeds. New fish hatcheries are being built. All wildlife ranging from wolves to wild flowers are now under complete protection.

By 1970 the seal population had increased to 45,000 and even the very rare Barguzin sable was staging a remarkable comeback."

The struggle to end the pollution around Lake Baikal is an example of both the difficulties and the seriousness with which a socialist government approaches this problem. The Soviet government issued its basic directives for cleaning the lake in 1969. Evidently the directives were not carried out fast enough. On September 24, 1971, both the Central Committee of the Communist Party and the Soviet Government issued directives demanding speedier action and full compliance. The new directives called for "speed in drafting and implementing the organization of the protected zone." The protected zone includes the drainage basin, in which are situated the mining and timber lands in the Lake Baikal region. The old pulp mills were given until 1972 to establish full pollution controls. The new pulp mills were instructed not to start operations until "appropriate treatment facilities were ready."

The directives set 1973 as the year by which the cities along the rivers that flow into Lake Baikal were to have full waste treatment devices. Electric power stations are prohibited from raising the water temperature in the rivers on which they operate by more than seven degrees fahrenheit.

Compare this to the futile efforts to save dying Lake Erie. Lake Erie is now listed as dead. It is not being cleaned up. If all of the pollution were stopped today it would take fifty years to give it new life. It took fifty years to kill it. In 1920 the commercial catch from Lake Erie was 33 million pounds of white fish, blue pike, and lake trout. By 1960 commercial fishing had to close shop. You swim in Lake Erie at your own great risk.

These are stories of two social systems and two lakes —one decaying and dying, the other flourishing and confidently looking to the future.

When the very first piece of evidence of the negative effects of DDT appeared, its production and use was totally banned in the socialist countries. In socialism there are no private corporations which, because of profits, would lobby against the ban.

The workers in our steel towns will understand the significance of the fact that 95% of all furnace dust from open hearth furnaces is captured by special filters in socialist countries.

The basic difference between the two systems is that in the socialist countries the battle for a livable environment is being won; the tide has been turned.*

* Gus Hall, *Ecology: Can We Survive Under Capitalism?* (New York: International Publishers, 1972). pp. 82-85.

part 4

- **The Struggle**
- **The Policy**
- **The Victory**

12 • National Liberation and Oil

The self-interests of the oil producing countries and the self-interests of the people of the United States are not in contradiction. Whatever differences do exist can be resolved on the basis of friendship and equality.

The struggle against domination of the seven corporate thieves, whether at home or overseas, is a struggle against a single foe. The obstacle to a reasonable price at the well head and at the gas pump is the profits of the oil corporations. Thus, the struggle against the Seven Thieves unites the people of the oil producing countries and the people of the oil consuming countries.

All the countries of the world have suffered from the profiteering of Standard Oil, Royal Dutch Shell, and their imperialist cartel partners. Repeatedly over the past century the people of producing countries have resisted and revolted against colonialism and neocolonialism; against the monopolies plundering their wealth, and against their local agents.

Decisive victories were won through socialist revolutions. The pioneering Russian Revolution ended forever the foreign ownership of oil and coal in the USSR. The oil and coal resources now belong to the peoples of Azerbaijan, Siberia, Russia, and the Ukraine, and to their many brother and sister peoples of the USSR. That takeover is irreversible. Repeatedly, ever since the 1917 Revolution, the imperialists have tried to get back that oil, stopping at no atrocity, no invasion, no intensity of economic warfare. But they have never gotten it back and never will—and now they know it. That was a qualitative turning point in history.

Socialist revolutions in Romania and Hungary similarly took away forever the oil fields the cartels had exploited before World War II and continued to operate willingly for Hitler during the war. The U.S. government raged, threatened, and refused to consider reasonable compensation, stole the Hungarian crown jewels, imposed embargos, but the cartels will never again appropriate the oil of these socialist countries.

After the overthrow of Batista, the Cuban government purchased Soviet crude oil as a normal business transaction. Standard Oil, which owned the refineries of Cuba, refused to refine it. This left the Cuban government no alternative but to nationalize the refineries. U.S. imperialism launched the Bay of Pigs invasion and provoked the "Cuban Missile Crisis" in a desperate, reckless attempt to retrieve its monopoly of Cuba's oil supply, as well as the nickel mines and sugar mills of Cuba for Freeport Sulphur and United Fruit corporations. But socialist Cuba will never yield these properties to the imperialist monopolies. And sensible people in the United States have come to recognize this.

In 1938, when the foreign oil companies refused to abide by Mexican law, the Mexican Government, headed by Lazaro Cardenas, nationalized the oil industry. U.S. imperialism, faced with war threats from the Hitler-Mussolini fascist axis, feared to invade Mexico; instead it engaged—unsuccessfully—in prolonged economic warfare against PEMEX, the Mexican government oil enterprise. But PEMEX has developed successfully; the first victory of a nonsocialist country against the oil cartel. But for a long time all other attempts failed.

In 1951 a progressive government in Iran nationalized the British-owned Iranian Oil Company. The imperialist countries successfully embargoed Iranian oil, and in 1953 the CIA was able to organize a military coup which overthrew the elected Mossadegh government, and installed a reactionary monarchy. And that regime, still in power, returned the oil to the cartel, and American companies were rewarded with a 40% share for the decisive role of U.S. imperialism in recapturing the oil.

With their domination in Venezuela and Canada, their takeover in the Middle East and the 40% control in Iran, the Rockefellers and the other U.S. oil pirates became the world's leading imperialist gangsters. From their plush Wall Street offices they dictated what was to be paid on the "barrel head" and set the wage levels in all the oil fields in the capitalist world. In a conspiracy with the U.S. government they shipped the oil into the country, paying only nominal tariffs. All this in the name of leading the "free world," which was free only to the oil hijackers.

With each barrel of oil the ambitions and the economic appetites of the bandits grew. The big business newspapers and magazines began to write about the "American Century," the Pentagon began to organize "paramilitary" units to operate with the CIA, and the generals began to talk about tactical nuclear weapons. And, as it was with German fascism, the smokescreen behind which all these activities were carried out was the big lie of anticommunism.

Imperialist Schemes Are Doomed

But life has its own laws of development. In the long run it does not bend to the schemes of profit-seeking marauders. The days of colonial empires and empire building have come to an end. A new world is "aborning." The oil empires are facing new and serious challenges. As a result the world balance of forces has dramatically shifted. The balance has shifted markedly in favor of the producing countries and against the oil cartels.

The breakup of the colonial system in the 1950s and early 1960s weakened the political positions of imperialism. Scores of new states appeared on the map, striving in varying degrees for an independent position, economically as well as politically.

The Soviet Union and other socialist countries achieved unprecedented economic and scientific-technical success. Increasingly they were able to help the newly free countries develop their own industries, providing machinery and equipment, generous assistance in training technical cadres, and markets for the materials of the newly formed manufacturing industries.

Countries that had no possibilities of obtaining modern economic development on their own, now had access to such development through economic cooperation with the socialist countries. Moreover, the Soviet Union now had sufficient military strength and material resources not only to defend itself and its socialist allies, but to provide weapons and other military assistance to developing countries seeking to defend themselves against imperialist aggression.

Within the developing countries a modern working class was coming into existence. Large numbers of workers, intellectuals, professionals, and military people by visiting and studying in socialist lands saw what socialism could accomplish. In many developing countries governments with an ultimate socialist orientation came to power. And in others, with conservative governments, the new social forces exerted considerable pressure on policies—pressure in an anti-imperialist direction.

The shift in the world balance of forces became the basis for a global, anti-imperialist front as expressed in the very first article of the International Covenants on Human Rights and on Social and Economic Rights, adopted by the United Nations in 1966:

> All peoples may, for their own ends, freely dispose of their natural wealth and resources without prejudice to any obligations arising out of international economic cooperation, based upon the principle of mutual benefit and international law. In no case may a people be deprived of its own means of subsistence.

These Covenants have been ratified by many countries, including the USSR. But the United States government has so far refused to ratify them because the monopoly corporations do not favor all peoples to "freely dispose of their natural wealth."

Since then many United Nations resolutions have amplified that general statement. In 1973 the General Assembly voted for a resolution which:

> *Strongly affirms* the inalienable rights of States to permanent sovereignty over all their natural resources ....
>
> *Supports* resolutely the efforts of the developing countries and of the peoples of the territories under colonial and racial domination and foreign occupation in their struggle to regain effective control over their natural resources;
>
> *Affirms* . . . that each State is entitled to determine the amount of possible compensation and the mode of payment (in the event of nationalization) and that any disputes . . . should be settled in accordance with the national legislation of each State carrying out such measures;
>
> *Deplores* acts of States which use force, armed aggression, economic coercion or any other illegal or improper means

in resolving disputes concerning the exercise of the sovereign rights mentioned in paragraphs 1 and 2.*

The resolution was carried 108 to 1 (United Kingdom), with 16 abstentions, which included the United States and most other industrialized capitalist countries. However, in the discussion the United States showed its actual opposition, and voted against key paragraphs in separate votes. Moreover, a separate resolution calling for Israel to cease stealing the resources of occupied Arab lands and the right of compensation to the people whose territory it occupied, was adopted by a vote of 93 in favor, 4 against. The four were: Israel, Nicaragua, Portugal, United States.** Some company!

OPEC Shakes the Seven Thieves

An important stage in the struggle against the oil cartel was the formation in 1960 of the Organization of Petroleum Exporting Countries (OPEC), which includes the main oil producing countries of Asia, Africa, and Latin America. It functioned as a collective bargaining agency of the producing countries and held regular negotiations with the representatives of the Seven Thieves cartel. At first OPEC bargained timidly for very minor concessions. By 1962 it succeeded in stopping the downward trend in prices the cartel was paying producing countries. Increases it negotiated during the remainder of the decade were insufficient to compensate for the rising costs of imported products, owing to inflation in the capitalist world.

However, in the 1970s there was a sharp increase in the militancy and objectives of OPEC. Setting the

* United Nations Document A/C.2/L.1328/Rev 1, and 1334; 4 December 1973.

** United Nations Press Release GA/4933, 17 December 1973.

stage for this were a number of new factors, in addition to the general ones already mentioned:

- The successful resistance of the Vietnamese people to U.S. aggression, which inspired national liberation forces everywhere.
- The growth of liberation struggles and radicalization of the Arab peoples, and their sharpened identification of friends and foes, following the Israeli aggression of 1967.
- Progressive governmental changes in Iraq and Syria and the winning of independence by Southern Yemen under progressive leadership.
- The increased dependence of the industrialized capitalist countries on oil from the producing countries, and the sharpening contradictions between capitalist countries, consumers of oil, and the Anglo-American-Dutch cartel.

With this new balance of forces OPEC exacted from the cartel substantial increases and other improvements in terms for the five-year period 1971-75. A further increase was imposed by OPEC at the start of 1972 to compensate for the devaluation of the dollar, and a similar step was taken in 1973.

More important, the producing countries moved decisively to take control of their oil from the cartel. In the late 1960s Bolivia and Peru nationalized the oil holdings of Gulf and Exxon, respectively. They were followed in 1971 by Algeria, and in 1972 by Iraq and Libya; all of which nationalized important sections of the foreign oil properties. By then almost all the main oil producing countries had set up national oil companies, had acquired a degree of knowhow of the world oil markets, had trained some oil engineers, technicians, and skilled workers, and had taken at least initial steps in the oil business. Through OPEC, the Persian Gulf producing countries announced that they would take a 25% ownership share in the oil fields in 1973, and that this would be increased to a

100% share in the early 1980s. The oil companies were forced to accept these terms in return for certain compensation and supply guarantees. Included in this arrangement were the very largest of the producers, Saudi Arabia and other oil states previously regarded as safe strongholds of imperialism under reactionary local rulers. Venezuela, Nigeria, and Iran also moved toward ownership and control, with some variations in the formula.

This led to a decisive sharpening of the contradiction between the producing countries and the imperialist cartel. The OPEC members moved in 1973 on the price front and insisted on getting something like reasonable payments for their oil, rather than prices that left them with only 10% of what the oil finally sold for to the ultimate consumers.

For the first time OPEC negotiations with the cartel failed to end in agreement. In September the talks broke down and the producing countries unilaterally declared an approximate doubling of prices to be paid them—to $3.05 per barrel at Persian Gulf ports, and somewhat more elsewhere. This price, about seven cents per gallon, was still well under the price of crude oil in the United States.

The biggest fear of the oil monopolies was coming to pass. The potential threat to their monopoly of the capitalist world's oil had become a reality, and with it a threat to their many billions of annual superprofits and to their objectives of infinitely expanding wealth and power.

The whole strategy of imperialism has always been directed toward preventing this through economic, political and military pressure. And it is inconceivable that the oil monopolies would peaceably and calmly now permit control to pass from their hands.

But the tactics used against Iran in the 1950s could not be used again. Boycotts were tried against Libyan and Iraqi oil after their partial nationalizations, but soon collapsed. The demand-supply situation in the world oil market, and the lack of unity among the capitalist countries make an effective boycott against even a single country unsuccessful. Against any large group of producing countries a boycott is absolutely hopeless.

Though the Seven Thieves hold most of the world's tanker fleet, their monopoly of transport capacity was being curtailed. About half of all vessels launched in the capitalist world are produced in the great Japanese shipyards, and most consist of supertankers, with deadweight tonnage in the hundreds of thousands of tons each. The Japanese are the largest competitors of the United States in the scramble for oil. Presently they depend, to the extent of 70%, on U.S. companies for deliveries. But their huge shipbuilding capacity gives them a degree of independence, a potential for breaking free of the grip of the Seven Thieves.

Deprived of effective economic weapons, imperialism turned to its principal weapon in the Middle East, political-military pressure.

Gendarme For Oil Imperialism

Since the middle 1950s the state of Israel has been the main local instrument for exerting that pressure against the Arab countries. Arising in a conflict between Zionist-expansionist elements controlling the emerging Israeli state and feudal Arab leaders, not yet displaced by the still weak national liberation movements, the Israeli government lined up with the major imperialist powers in oppositon to the Arab liberation movement as it developed. Financed by billions in donations from the United States and other capitalist countries, armed to the teeth, and supplied with technically skilled manpower by the imperialists, it sought to become a junior regional imperialist power, expanding in territory and resources at the expense of Arab neighbors.

Israel and the imperialists coordinate these efforts to impede the progressive tendencies in the Arab countries. Thus, after the Egyptian Revolution of 1952, which culminated in the expulsion of British troops and the nationalization of the Suez Canal in 1956, the Israelis attacked Egypt in conjunction with British and French armed forces. Thanks to the Soviet Union's firm support of Egypt and contradictions between U.S. and British imperialism, the invaders were forced to withdraw.

However, by 1967 the U.S. government had become the main direct supplier of the Israeli military, and the leading instigator of Israeli aggression. This time Israel launched an aggressive assault against Egypt, Syria, and Jordan, seizing big chunks of territory from each of these countries, appropriating an area twice its own size. Again Israel was stopped from further advances only by the threat of Soviet armed intervention. But, with the support of U.S. imperialism, Israel persistently defied the United Nations Resolution 242 ordering its withdrawal from the occupied territories and guaranteeing its security as a state. Instead, it has tried over the years to gradually settle these territories, fortify them, and exploit their natural resources.

As the conflict between the oil producing countries and the monopolies reached a climax, Israeli provocations against its neighbors accelerated. There were massive aerial assaults on Syrian seaports and airports, punitive invasions of Lebanese territory, and the invasion of Beirut, the capital of Lebanon by Israeli armed forces to assassinate leaders of the Palestine Liberation forces.

The United States government vetoed urgent measures of the United Nations Security Council to head off another military showdown, to force a change in Israeli aggressiveness. The United States government was warned by the Arab nations that this course would make war inevitable. The oil companies were warned about the certainty of an oil embargo and the likelihood of a showdown in the Middle East between the producing countries and the monopolies.

The basic policy line of imperialism was to create and use the energy crisis to preserve control of the oil, and to keep the whole area of the Middle East firmly under the heel of U.S. and Israeli imperialism.

This is illustrated by the penetration of finance capital into the Middle East. *Business Week*, 16 March 1974, reports that "U.S. banks are opening new branches in the area and buying into local commercial banks, and they are setting up joint-venture merchant banks with Arab partners as well." In fact this magazine finds that they are "combing the area so intensively for business they are practically bumping into each other."

First National City Bank of New York has branches in Bahrain, Dubai, Abu Dhabi, and Qatar, and two in Saudi Arabia. Bank of America has had a branch in Beirut for many years and now is expanding in the Middle East chiefly through its share in the Bank of Credit and Commerce International, which it established in Luxembourg two years ago. The American Express Middle East Development Company has been set up recently by American Express. Chemical Bank last year bought 80% of Beirut's Rubiya Bank, and Irving Trust is negotiating to take over a bank there. Chase Manhattan is also very active in the area. These are only a few examples.

The October War of 1973 showed how drastic were the changes in the world balance of forces. Syria and Egypt had reformed their armed forces, acquired better military technique and morale, and benefited from new and superior Soviet weapons and training. They fought the Israelis to a standstill, and destroyed the myth of Israeli invincibility. The October war was a historic setback for U.S. imperialism.

U.S. imperialism was isolated as the sole major supporter of Israeli aggression. All its allies refused to permit use of their territory for the delivery of supplies, except colonialist Portugal. United States attempts to put together a "consumers union" of industrialized capitalist countries to confront OPEC collapsed as the other capitalist countries seized the opportunity to deal directly with producing countries rather than through the U.S.-British-Dutch cartel.

The Arab countries showed a much higher degree of unity and carried out a curtailment of output and an embargo on shipments to the United States more successfully than formerly. More important, they used oil money to finance the heavy expenses of the war for Egypt and Syria.

The Middle East crisis continues. The exact outcome of peace negotiations remains to be seen. But events are moving inevitably toward a resolution of the long conflict:

• The Israeli aggressors will finally have to withdraw from conquered territory, as the U.S. invaders were finally forced to withdraw from their bases in Indochina.

• The progressive forces in the Arab countries will be strengthened and these countries' relations with the Soviet Union and other socialist countries deepened.

• The process of complete nationalization of oil has been accelerated. It now seems likely that before the end of 1974 national oil companies will have majority or total ownership of the oil in all of the biggest Middle Eastern countries, including Saudi Arabia, Iran, and Kuwait, as well as the more progressive regimes. The cartel members will still be able to buy a large share of the oil, but on terms dictated by the producers.

• Consuming countries, such as England, France, Japan, and Italy, are hastening to make direct deals for supply with the producing countries, ignoring the cartel. In return, they are agreeing to supply capital, machinery, and knowhow for industrialization and technical progress.

The producing countries are no longer satisfied with getting payment in Swiss bank accounts to pay for the entertainment and luxury of narrow, ruling cliques. Learning from the mutually beneficial relations of developing countries with socialist countries they are demanding that in exchange for vital fuel the consumers of oil provide what is necessary for developing the productive machinery and general level of life in the producing countries. And the capitalist countries, having no other choice, are accepting these terms.

The "solidarity" of the imperialist states against the national liberation movement is collapsing as governments find that U.S. imperialism can no longer take care of their material needs in exchange for political support. The imperialist oil cartel declared nationalized Libyan oil "hot oil" that no capitalist country was expected to handle. The socialist countries ignored the dictates of the oil cartel. And following an internal struggle within the Japanese ruling circles, Japanese companies were authorized to purchase the "hot oil."

The Arab countries embargoed oil shipments to South African, Rhodesian, and Portuguese colonialists and racists, establishing a new level of solidarity with the other countries south of the Sahara. These countries, in turn, broke off relations with Israel, nationalized oil installations, and made direct deals with producing countries for their supply of oil in exchange for their products.

Formerly only the developing countries with progressive governments would take serious action against imperialist monopolies. But today such actions are taken even by states (Saudi Arabia is a case in point) where the progressive forces have not yet obtained power, and where reactionaries still hold the reins, but have to respond to the changing world conditions and the growing pressures of progressive groups within their own countries.

Also instructive are the actions of the government of Zaire (formerly the Belgian Congo). Following a visit of President Mobutu to Algeria, Libya, and Egypt, Zaire has taken over control of all oil tanks, pipelines and tankers of foreign oil companies operating on its territory.Zaire was guaranteed a direct supply of oil without going through the imperialist cartel, and the nationally controlled installations will be used for the transportation and storage of Arab oil. The action was taken despite U.S. and Israeli threats to cut off "aid" if Zaire continued to support the Arab cause, and despite the threats of the monopolies to deprive the country of oil. It is significant that Zaire has a capitalist government whose president came to power as a tool of U.S. imperialism and acted as executioner of the heroic revolutionary Lumumba.

But even such a person as Mobutu, and the young bureaucratic capitalist class built up around his regime, cannot ignore the changing balance of forces; cannot fail to see the advantages of breaking from the domination of the imperialist oil cartel.

A major objective of U.S. imperialism, repeatedly expressed, is to drive a wedge between the Middle Eastern countries and the Soviet Union, to prevent the increase of Soviet influence in the area. But the opposite has occurred. The consistent principled position of the socialist countries has resulted in expanding and deepening relations with the developing countries.

Soviet planes flew over Yugoslav territory to deliver vital armaments to Egypt and Syria. As "punishment," the U.S. oil companies broke their contracts to deliver oil to Yugoslavia. Algeria sent Yugoslavia supplies to meet the emergency, and in a long-term agreement Libya guaranteed steady shipments of oil to Yugoslavia. In turn, Yugoslavia agreed to assist in a number of development projects in Libya.

This is the trend of events in the Middle East and it is not promising for imperialism. The scenario may change; there may be setbacks. But given the existing balance of world forces, the direction of events cannot be changed. And any attempt at extreme adventurism on the part of the United States and Israel would be to risk a global thermonuclear catastrophe. Of course, the best insurance against such adventurism is the development within the United States of the same understanding of the issues of the Middle East and oil as led to the great movement against U.S. aggression in Vietnam.

What is happening in oil is not unique. In copper, in coffee, in bauxite, in other raw products, the developing countries are uniting their policies taking over ownership and control, setting prices for sale to the imperialist countries. The period when imperialist countries can get materials at the "cost of extraction" is coming to its inglorious end.

The present struggle over oil is part of the thrashings-about of imperialism, striving to hold back the tide of the world movement for national liberation and socialism. The interests of the American people are not those of the oil trust, nor of the administration, the Congress, or the media which supports the aims of the oil cartel. Our interests are on the side of the world's socialist and national liberation forces. Our people inevitably face a similar struggle on the homefront against the same monopolies. And in the meantime we have the duty, in our own best self-interest, to stop their aggressive actions abroad.

OPEC is denounced as a "cartel," and the consuming countries are urged to form a "countercartel," to defend themselves against OPEC's "exactions." But the whole comparison is wrong. OPEC is an organization formed to defend the producers *against* the real imperialist cartel of the oil monopolies, to negotiate the terms on which they will sell their irreplaceable natural resources to foreign users.

The assumption of the U.S. press, and that of most imperialist countries, is that the producing countries have a *duty* to supply oil to the imperialist centers. *But they have no such duty.* The oil cartel has, in some cases, taken out profits a hundred times greater than the amount invested in production. The Arab countries do not owe oil to the United States, Japan, or Britain, any more than the United States owes machinery or tobacco to some other country.

If a country wants to buy oil from a producer, they can bargain on terms of equality, and agree to supply what the producers want and need in exchange. And that is exactly what is happening as the grip of the oil cartel begins to break up. And it is to the long-run advantage of the people of the consuming countries as well as those of the producing countries. It means that oil producing countries will get equipment and knowhow needed for their development, as well as helping to end the age-old poverty and ignorance of the masses. Meanwhile, for the workers of the oil consuming countries, it means jobs producing machinery and industrial supplies for the producing countries.

Senators Henry Jackson and James Buckley, the B'nai B'rith and other Zionist groups, the *New York Times*, and imperialist spokesmen generally, denounced the "Arab oil blackmail" when the Arab countries imposed a mild reduction in output and embargoed shipments to the United States, the principal supporter of Israel. It is interesting to watch these contemptible phonies and even the disbelief by some that the United States should be getting the other end of the stick in the very game it invented. That game has been a central pillar of U.S. foreign policy for at least fifty years. It is called "economic blockade and embargo," and has been an exclusive U.S. weapon, used especially in the struggle against the socialist countries and the countries that have won independence. It was used against the democratic government of Chile. It is the main weapon, to this day, against Cuba, to whom the United States still refuses to sell anything and from whom it still refuses to buy anything. And the United States still maintains an embargo policy against the People's Republic of Korea, the Democratic Republic of Vietnam, Laos, etc.

So now the hypocrites, the doubledealers, people with no moral standards who supported the U.S. imperialist policies of embargo, are screeching from the housetops against the Arab peoples. When the United States used this policy as an imperialist weapon it was acceptable, but when the Arab people use the very same policy it is called blackmail.

Then there is the hypocrisy of the eight economists, four of whom have won Nobel prizes in economics, including Samuelson and Galbraith, leaders of the "liberal" U.S. economists. They emerged from their economic mothballs to proclaim: "Our foreign policy should not be deflected in the slightest by illusions that giving in on oil blackmail will in fact gain us anything." But these economists, glorified by the bourgeois establishment, never lifted their voices against the embargoes carried out by U.S. imperialism, and still do not. Perhaps it is understandable why the Nobel prizes are given to so many spokesmen of imperialism. Nobel was not only a producer of gunpowder for imperialist wars, he was also the first big capitalist to lose a major property to the world's revolutionary forces when the Russian Revolution nationalized his Baku oil properties.

There is a serious side to this campaign about "Arab

blackmail." It is accompanied by open threats of invasion of Arab lands to get the oil. The United States has demonstratively sent a fleet into the Indian Ocean to threaten the Persian Gulf lands, just as it keeps carriers in the Gulf of Tonkin to threaten Vietnam. Those who say the United States has a "right" to Arab oil are claiming that it has the "right" to follow in the footsteps of Teddy Roosevelt and Woodrow Wilson and march in to grab the oil.

Well, Times Have Changed

This threat of invasion is a most dangerous game today, and the American people must stop it in time—not only because an imperialist invasion is unjust but in our own interests. Such an imperialist invasion, in the hotbed of the Middle East, could spark World War III and destroy all of us.

The process of nationalization is also a law-governed process that will continue in one form or another. As a fact of history, colonialism is on its way out. Neocolonialism cannot replace it. What we are witnessing is the passing of oil colonialism. The world is catching up with the Seven Thieves.

The share of government production of oil in the developing countries increased from 8% in 1970 to 11% in 1972; and for the Arab countries alone the increase was from 3% to 10%.

During this time the relative share of the imperialist powers in the developing countries dropped by 3% and in the Arab countries by 7%. True, this amount is considerably less than the volume of oil still being produced and exported by the Seven Thieves in the area. However, the success of the National Liberation Movement in the struggle for oil is irrepressible, and even more rapid gains have been made since 1972. A few examples are in order to illustrate how fast revolutionary developments are moving.

On January 3, 1974, the Kuwait government announced its agreement with two of the Seven Thieves—Gulf and British Petroleum—which calls for a 60% takeover from these two giants that had set up the Kuwait Oil Company in cloak and dagger fashion. Maintaining a facade of national respectability, the company had been owned lock, stock, and barrel by these two thieves, who behaved as if they owned not only the oil resources, but Kuwait itself, including the people.

But with the government takeover all that has now changed. The agreement also provides for total nationalization of the company by 1979 if approved by the Kuwait parliament which historically has been a major force for nationalization. Kuwait's oil reserves of 65 billion barrels in 1972, together with Iran's of the same amount, trail only Saudi Arabia's 138 billion barrels for supremacy in the Middle East.

In September of 1973, Libya, the world's seventh largest oil producer, passed a decree for the nationalization of 51% of all foreign oil companies operating on Libyan territory. This included 51% of Exxon's wholly owned Esso Libya.

On February 11, 1974, Libya announced it was completely nationalizing three American oil companies that account for 5% of Libya's daily output. Incidentally, the announcement was made on the opening day of the abortive U.S.-sponsored conference of major industrial capitalist powers, and was an effective warning that imperialist threats will not work. The conference had been set up to torpedo the anti-imperialist initiative taken by the Middle Eastern nations, including Algeria's call for a United Nations special session to discuss the question of trade—raw materials—including oil, on a worldwide basis. The capitalist conference was abortive because, among other reasons, even leading imperialist nations such as France refused to go along.

Plans are under way for the nationalization of the claimed $1.5 billion U.S. petroleum investment in oil rigs, office buildings, refineries, pipelines, and other facilities in Venezuela, the largest supplier of oil to the United States. Here the largest foreign monopoly is Exxon's subsidiary Creole Petroleum. Present plans call for the nationalization of foreign natural gas holdings when the concession agreements with the Venezuelan government expire in 1983-84 and 1996-97. Companies have been required to post bonds totaling more than $500 million to insure that they keep their plants in good condition for delivery to the government.

These three countries alone—Kuwait, Libya, and Venezuela—accounted for one-sixth of world oil production in 1972. Kuwait and Venezuela are among the exclusive group of six nations that annually produce more than one billion barrels of oil each. And Libya is a close seventh, producing 810 million barrels in that year.

In February 1971 Algeria nationalized the controlling interests of the foreign concession on its territory, and as a result Algeria's share in the production of her oil increased to 80%.

In June 1972 the socialist-oriented Iraq Republic took over the largest of the three branches of the Iraq Petroleum Company that operated on its territory. By the end of 1973 almost all remaining foreign-owned oil was nationalized. Large new oilfields are being developed with Soviet assistance.

In 1973 Saudi Arabia demanded and received a 25% interest in the Rockefeller-dominated ARAMCO, which operates the world's most productive oil concession. The agreement called for increasing Saudi Arabia's share each year to a maximum of 51% by 1982.

Similar arrangements were made byAbu Dhabi and other producer nations of the Arabian Peninsula. Now, however, especially with Kuwait's imminent takeover of 60% of the holdings of the major foreign monopolies, even the ARAMCO settlement—an event of great significance in its time—is outdated.

Today there is far greater pressure on the Saudi Arabian government to fully nationalize rather than remain in "participation" or "partnership" with the imperialists.

There are currently two opposing tendencies in the Middle East concerning nationalization. The more reactionary regimes led by Saudi Arabia favor participation with imperialism.

"We called for participation," said Sheikh Ahmed Zaki Yamani, the Saudi Arabian Minister of Oil and Mineral Wealth, at a conference in London in September 1972, "because the alternative which is nationalization conflicts with our way of thinking and our economic system."*

On the other hand, the countries with progressive regimes, notably Syria, Algeria, and Iraq have different views concerning the idea of partnership. Iraq and Algeria are the tenth and thirteenth largest oil producing nations, respectively. Nationalization is an inevitable step toward economic and political independence. The partnership and participation scheme is neocolonialism.

At the oil seminar held in Baghdad in November 1972 on the initiative of the World Peace Council, the Afro-Asian People's Solidarity Organization, and the National Council for Peace and Solidarity in Iraq, the representatives of these and other countries unanimously rejected the partnership system, assessing it as a step backward compared with nationalization.

The rejection of partnership and participation is

* *International Affairs*, May 1973, p. 50.

part of the process of the changing balance of world forces against U.S. imperialism. This has been admitted by organs that normally support the establishment. A study by Mathematica, Inc., a Princeton, New Jersey research firm concluded: "There can be no doubt that the coming world energy squeeze will tend to result in a vast redistribution of international power and influence—to the detriment of the United States. . . ."*

This illustrates a central theme of the Conference of Communist and Workers' Parties held in Moscow in 1969: "*Imperialism can neither regain its lost historical initiative nor reverse world development.*"** Imperialism cannot return to gunboat diplomacy. It cannot "deal from a position of strength." It is forced to accept nationalization and independence.

The happenings in oil are an accurate mirror of what is happening in the field of raw materials generally. A hundred-year history is fading out, bringing to an end a century of corporate-imperialist robbery. The world is no longer accepting colonialism's slavery in any form.

The United States is self-sufficient in only ten of the thirty-six basic industrial raw materials. Following the example originally set by U.S. and other monopolies, more and more countries may boycott the U.S.. robber barons and demand the end of neocolonial domination.

The time of colonial oil is gone.

The days of U.S. foreign raw-material domination are numbered.

The robber barons and their spokesmen are well aware of this fact. Thus, Arthur Burns, chairman of the Federal Reserve Board, said:

* *New York Times*, 14 November 1973.

** Documents adopted by the International Conference of Communist and Workers' Parties (Moscow, June 5-17, 1969, p. 14 (emphasis in original).

> Recent events are very disturbing . . . . What happened in oil could happen [in copper and other raw materials. There is] no reason why the group of countries that supply most of our bauxite [the ore from which aluminum is produced] can't get together the way the [oil producing] countries got together on the price of oil.*

Of course it is "disturbing" to the corporations —but it need not be disturbing to the people of the United States.

In addition to OPEC there are other economic organizations involving leading nations in the national liberation movement. These include the Intergovernmental Council of Copper Exporting Countries (Chile, Peru, Zambia, and Zaire) and the International Tin Council (Malaysia, Bolivia, Indonesia, Nigeria, Zaire, and Australia).

Raw materials will continue to reach the United States, but the prices will reflect the supply and demand. The new prices will be less and less influenced by colonialism.

13 • The Path Ahead—Nationalization

The energy crisis may be contrived but it hits the working people with the full force of the real thing. For most Americans there is a real energy crisis, a shortage of available energy and a shortage of money to pay the soaring cost of energy.

That there is really plenty of potential energy available, that it really remains relatively inexpensive to produce doesn't help the workers who cannot get sufficient gasoline for their cars, or who have lost their jobs on the basis of the energy crisis, and who pay exorbitant prices for gas and fuel oil because of corporate profits.

* *Wall Street Journal*, 26 December 1973.

To see the way out let's examine in more fundamental terms the solutions offered by big business and the Nixon administration.

On the domestic front their formula is quite simple: Let prices go high enough, then only the rich will be able to buy; there will be a greater supply of food, gasoline, and fuel because the poor will not be able to afford them. Corporations can raise prices, sell less, and make bigger profits.

But that is no solution for the people or for the country. The obvious flaw is: What's the use of having plenty of energy again if a large part of the population cannot afford to buy it and the wild inflation of energy prices follows the still continuing inflation of food prices? The higher prices of energy, even more than of food, are "spilling over" to scores of other commodities, infiltrating the entire economic structure and generating runaway inflation all along the line. A corollary is that the Nixon formula is bound to increase the instability of the economy, to plunge the country into worse financial crisis, worse recessions and depressions, more serious and long-lasting bouts of very high rates of unemployment. Again, what's the use of having plentiful energy supplies if the economic conditions make their consumption impossible?

But that isn't all. It isn't part of the big business game plan to provide plenty of energy, even given higher prices. Indeed, all of their spokesmen, from energy czar Simon to President Nixon to the chiefs of the oil companies, stress the theme that we are going to have to live with shortages for a long time to come. They suggest fifteen years! They say we are going to have to change our life-styles and so on. Nixon has called on the American people to "develop a national energy conservation ethic" and to "sacrifice" to alleviate the energy crisis.*

* Speeches, 19 April 1973; 20 January 1974.

Needless to say the millionaires do not intend to change *their* style of life, except to make it more luxurious. But for the rest of us this means to worsen our style of life, to reduce our standard of living because there will be no improvements in public services, in planned coordination of housing and industrial location, in ending discrimination that would make practical and desirable a reduced use of private automobiles.

But notice the contradiction between the two themes: Grant higher prices and there will be plenty of energy, on the one hand—and, prepare for a long siege of shortages, on the other. The fact is that the real game plan of the monopolies is to organize and maintain permanent shortages, to create the environment for maintaining very high prices and periodically raising them. From this point on the inflationary tendency is going to be a continuous factor in capitalist development.

To see what the game plan projects for us, just look at the public transit system of our cities. Fares have multiplied five to ten times since World War II. And as service deteriorates and becomes more limited, the number of passengers is reduced. The combination of decreasing transit services in relation to needs and multiplying monopoly prices is exactly what the energy monopolies have in mind. And they mean to organize it and enforce it with the aid of the state.

On the foreign front their approach is similar in intent, if somewhat more obscure in statement.

The continuing drive of Nixon and Kissinger is to organize a united front of consuming countries —actually a united imperialist front of the industrialized capitalist countries—to deal with the producing countries. The idea is to restore the earlier situation when the oil monopolies and the U.S.-British governments had such a stranglehold on the capitalist world and the producing countries were so divided that the imperialists could organize and enforce an embargo against the oil of any recalcitrant producing country.

They aim to get the consuming countries to isolate a chosen group of the producing countries and dictate terms to them on pain of an embargo on their products. The terms of course would be maintenance or restoration of effective control of the oil in the hands of the Seven Thieves, and a reduction in their payments for the oil.

And the aim isn't really well-hidden. Hinted in print and spoken in private is the plan to create a political situation enabling them to use NATO for an armed assault on selected producing countries if that seems necessary. This objective of economic and perhaps military warfare against the producing countries is the reason Nixon and Jackson unite in calling for achieving self-sufficiency in energy by 1980. It is not self-sufficiency they want, but sacrifices from the people on consumption of energy. This paves the way for them to use billions of tax dollars to develop new energy sources to put the monopolies in a stronger position to carry out their planned embargoes.

This is a futile and possibly fatal game. The interests of many other capitalist countries vary considerably from those of the U.S. imperialists. In particular, most of them have been victims of the U.S.-British-Dutch oil trust, and not partners in its profiteering. They vary greatly, too, in their dependence on imported energy; in the cost to them of any embargo campaign; in the level within each of the working class-led antimonopoly struggles that would resist any attempt to cooperate with this scheme of Washington. And finally, the plan reckons without the reserves for resistance now available to the producing countries: The advance in their own national liberation movements, their level of cooperation, and the massive support available to them from the socialist countries when they engage in a struggle against imperialism.

Cosmetic Changes Not Acceptable

The time for "cosmetic changes" has passed. History has placed basic revisions in world relationships on the order of the day. The imperialist countries can no longer dictate or "deal from a position of strength." The shift is to nationalization.

Cosmetic changes are not acceptable solutions to domestic problems either. Most of the proposals coming from Establishment sources are fakery. Sen. Jackson's and the Nixon administration's proposals for an "excess profits tax" as the solution is a smoke screen behind which the oil corporations would continue to escalate prices. If such a law were passed, a flick of the computer switch would "prove" that the corporations are "losing" money. Even if they admitted "excessive profits" they would not have to pay taxes, because they have built up a five-year reserve—a negative income tax which they can count on.

In a rare moment of truth an executive of an oil corporation stated the following to a reporter of the *Chicago Sun Times:*

> They don't care if there is an excess profit tax because they'll just decrease their profits by paper work.
>
> There isn't an accountant I've ever talked to at my company who knows what the hell is going on with their books.
>
> They send one set of books to the IRS and that's the same information that they are giving to the Senate committee. The way their accounting is set up, they could hide anything.
>
> They have a firm that certifies their accounting for the government. But they don't conduct a thorough audit. They

just certify the books. The IRS gets the books and they really don't know a thing.*

No. The Establishment has no solution for the people. Not only is it possible, but it is vital to find another way to meet the growing energy needs of our country and of the world. The time has passed when palliatives can help, when regulations by the Federal Power Commission or endless courthouse maneuverings by the Anti-Trust Division of the Justice Department can make a major difference.

There is only one way to deal with the problem in the interests of the people. And that is through nationalizing the entire energy complex, lock, stock, and barrel, and putting it under the control of a democratically elected National Energy Council.

Lee C. White, a former chairman of the Federal Power Commission, and Sen. Adlai E. Stevenson, III (Dem.-Ill.) favor creation of a federal corporation to explore for oil and natural gas. Edward Cowan of the *New York Times* writes of this proposal that: "It is an idea whose time, politically speaking, may be near."** But if the products and discoveries of such exploration go back to the oil monopolies it will solve nothing.

A bill before Congress proposes a government corporation to build refineries. Again, for the government to build refineries and then turn them over to the oil monopolies is no solution. The United Electrical Workers proposes "development under government ownership and control of gas and oil from shale."*** This is a step forward.

As Cowan of the *New York Times* points out, the criticism of the oil industry by Nader and other liberals "seems to point logically to nationalization," although many of the liberals, including Nader, haven't

* *Chicago Sun Times*, 6 September 1973.

** *New York Times*, 16 December 1973.

*** *U.E. News*, 3 December 1973.

gotten to the point of saying it. However, more and more people do, especially workers.

The energy crisis has not only exposed the contradictions of capitalism, but has greatly heightened the awareness of its decay. There is a sharp turn in mass patterns of thought. "Profits" has become a dirty word. Monopolies are seen as the villain by many. The shield of capitalist demagogy is wearing thin.

The energy crisis is forcing changes in attitudes and relationships and has its effects on class relations. There is a rising class consciousness in the ranks of the workers, a growing awareness that the crises of shortages means rip-offs for the capitalist class. In many ways this is adding a new dimension to the class struggle, which in turn is creating new problems for the servile top leadership of the trade union movement. The ideology and the policies of class collaboration are historical roadblocks to concepts of nationalization.

Class collaboration policies of trade union leaders are an old curse, an old millstone around the neck of the working class. There are of course important exceptions—labor leaders who do not play the corporate game of collaboration—but unfortunately they are as yet the exceptions to the rule.

Class collaborationism has always been fed by the objective factors in U.S. capitalist development. Cheap colonial raw materials, cost-of-extraction raw materials at home, an abundance of labor power, the extra profits from racist oppression of Blacks, Chicanos, Puerto Rican and American Indians have all created the objective base for these policies. When the capitalist world was Wall Street's oyster, class collaboration was its reflection in the trade union leadership—there were more crumbs to pass around.

Now the world is changing. The oyster beds are moving into deeper waters. A new world in which the oysters are not available for the asking is emerging.

The corporations now have to negotiate for the purchase of raw materials. "Purchase" is taking the place of "plunder."

But the corporations are not giving up their drive for maximizing profits. In oil we are beginning to see what is meant by the new life-style the corporate propagandists are talking about. "Passing on the cost-plus" to the consumer is the new life-style of capitalism.

It is becoming somewhat more difficult to talk about collaborating with a class that uses the energy crisis to lay off workers, a class that escalates prices, taxes, and rents, a class that uses "shortages" to squeeze out the biggest profits in the history of capitalism.

Throughout history the only obstacle to total sellout by the top leadership has been a militant rank and file.

The new stage of capitalism is forcing a new period in the class struggle. It is giving rise to a new movement of the rank and file that rejects the policies of class collaboration. Thus, these changes in the class struggle are the result of changes in the objective class relationships. These changes are clearly present in the resolutions for nationalization now being adopted by the trade union movement.

Radical Demands Are Surfacing

Joseph Mangina, business agent of Local 301, International Union of Electrical, Radio and Machine Workers, AFL-CIO, representing workers at the huge General Electric plant at Schenectady, writes:

> Congress who represents the interest of all Americans should assume the leadership during this crisis. Perhaps one of the first things they should consider is the advisability for the nationalization of oil, coal and natural gas industry, regulating the output and distribution for the duration.*

* *301 News*, 18 December 1973.

Nicholas Von Hoffman, *Washington Post* columnist, writes:

> Nationalizing the whole thing is the only sensible thing to do. If we must have a planned economy, then we, not the major oil companies, should do the planning and profiting. No socialist government entity will do worse.*

During most of its history the organized trade union movement has openly supported capitalism. This has been official policy. It has never fully reflected the sentiments of the rank and file. Antitrust, antimonopoly sentiments, concepts of nationalization, of public ownership, and of socialism have deep roots in the ranks of the membership.

The energy crisis has forced these sentiments to the surface. They are reflected in the statements and resolutions passed by many unions, including the Amalgamated Meat Cutters and Butcher Workmen, the Service Employees International Union, the American Federation of State, County and Municipal Employees, Automobile, Aerospace and Agricultural Implement Workers, and others.

Nationalization of the oil industry is also recommended by the Pacific Northwest Labor Advocate, official organ of the regional AFL-CIO Council and by the Minnesota Central Labor Union Council. The Massachusetts State Legislative Black Caucus has taken a similar position. (See Appendix C, p. 212 for a partial listing.)

Harry Bridges, the president of the International Longshoremen's and Warehousemen's Union, discusses the question in *The Dispatcher*:

> Various people have various ideas about the U.S.. Some say the oil industry should be nationalized. While this might seem in the view of many to be desirable, it is not very feasible. Industries in the U.S. are not nationalized easily.
>
> It has happened temporarily—in certain emergencies-

* Reprinted in *New York Times, 16 December 1973.*

—and even then the nationalization wasn't very real. The same management remained in control.

But *unions do* have power.

Just as the Arabs have asserted their power by withholding their oil for political reasons (and raising the price for economic reasons), so do workers have the right to assert their power to avoid being rendered jobless simply because some company wishes to build a pipeline or drill an offshore well or start mining shale.

If there is going to be any withholding, perhaps the unions should do some withholding of their labor power to make sure that the oil companies do not continue to manipulate this situation to their own advantage and to the disastrous disadvantage of working people. We have our weapons too.

This is, as I say, a great chance for the labor movement to show whether it is worth its salt.*

Bridges raises some very serious problems. He is correct—industries in the United States are *not* nationalized easily. The line of struggle he proposes, of workers "withholding their labor power," is not necessarily in contradiction to the demand for nationalization.

The idea of nationalization is not new. The steam that is now building behind it is new, and reflects the thinking of broader masses that to simply reform capitalism is not the answer. The idea of nationalization must be seen in the context of the new objective conditions that are forcing it to the surface.

The Policy of the Communist Party

Reflecting these new developments the National Council of the Communist Party, U.S.A., at its December 1973 meeting, adopted the following policy:

The nation cannot afford to have its life-and-death re-

* *The Dispatcher*, 11 January 1974.

sources in the hands of irresponsible, coldblooded gangsters without a social conscience. Public utilities must belong to the public.

There is need for a totally new energy structure. There is a need for a democratically elected and controlled National Energy Council—a council of people's power; a council that will own and operate all energy-related businesses from oil, coal, and gas fields to the refineries, to their distribution, including all power plants and nuclear plants. This National Energy Council would operate the energy-related establishment on a nonprofit basis as a real public utility. It is an idea whose time has come—it is an antimonopoly measure that can get the support of broad masses of people. We should combine this campaign with more immediate demands such as:

1. Build government-owned and -operated refineries.
2. Stop the flow of fuel to all military operations, including the production of munitions.
3. Take a forthright stand on U.N. Resolution 242; close the military pipeline to Israel, to South Vietnam, Cambodia, South Africa and Chile—and open the oil pipeline to the world.
4. No energy-crisis-related layoff of workers; demand that the corporations who created the crisis continue to pay the wages of the workers who are laid off.
5. Cut off the luxury fat of the rich, such as private planes, limousines, and yachts.

These are important demands. But only by taking over the energy complex from start to finish can there be a rational plan for the use and production of energy, with safeguards of conservation and avoidance of pollution.*

The report notes that:

. . . the immediate goals of the struggle must be a demand for federal laws—without the Nixon clauses that make them meaningless.

The demand must be for laws with provisions through which the people, workers, the people in the ghettos, trade unionists, can be enforcers.

* Gus Hall, *The High Crimes and Misdemeanors of Monopoly Capitalism* (New York: New Outlook Publishers, 1974), pp. 22-23.

The demand must be for the nationalization of industries that violate antipollution laws.

The demand must be for laws that guarantee that the cost of all antipollution measures comes from corporate profits.*

The key to unlocking the pollution crisis lies in the successful struggle around the immediate and long-term demands outlined in this policy.

The oil and power monopolies have been engaged in a criminal conspiracy against the people of this country. They have been guilty of extorting billions and billions in superprofits at our expense. The weapon of ownership of the nation's fuel and power apparatus should be taken away from them and made into a real nonprofit public utility—publicly owned, controlled, and operated.

What about compensation? In general, the main owners of the monopolies have enjoyed profits many times over their original investments; the Rockefellers, Mellons, Gettys, et al., are not entitled to any compensation. But this is a problem that can be settled in a reasonable way, with compensation to small stockholders and bondholders—subject to a flat upper limit to any individual's or family's compensation. The point is to avoid paying out so much in compensation that the former owners have command over a national tribute that nullifies the objectives of nationalization. That is what has happened with the British variety of nationalization of the coal and steel industries.

By a National Energy Council to run the energy industries we do not mean a body headed by some Nixon-appointed czar. We mean an elected body; and elected according to certain stipulations that assure its actual democratic character. Some members of the council should be elected by votes of all the workers in the energy industries, representing the interests of the producers of energy on the council. And some (a

* Ibid., p. 88.

smaller number) should be elected by the professional and technical specialists in the field of energy, including specialists in workers' health and safety and in environmental standards.

The National Energy Council should operate the energy industries on a nonprofit basis. Research and development should be financed by the federal government, out of taxes levied on the rich and the corporations.

Prices of oil, coal, gas, and electrical power should be rearranged to stop the gross discrimination against individual consumers in the monopoly price structure. Working people would be guaranteed ample supplies of energy at stable, low prices.

The development of new sources of energy should be organized according to a single, long-term plan, based on projections of scientific discoveries and their adaptation to economical, practical use, on projections of the increase in the requirements of energy and of the forms in which it will be needed. The plan should be coordinated with a nationally planned transportation system, and with housing and city planning, so as to dovetail the buildup of energy supplies with rational schemes for economy in its use and convenience in its enjoyment by the people.

It would be foolish to try to spell out all details of this proposal. It will have to be worked out in a democratic way, by the millions who will have to be involved in the struggle to realize it. For certainly it cannot be won without such a struggle. And yet we are convinced that the time is ripe for that struggle and with good chances of success in the foreseeable future.

This would be a demonstration of people's power at its finest under capitalism. It could put an end to the contrived energy shortages, the profiteering, including the use of oil depletion and "intangible" expense tax dodges, and the wasteful use of energy resources.

Great victories in the struggle for a living environment could also be won.

It is the only way that real gains can be made, short of socialism and a transfer of class power to the working people.

High on the list of energy savers, along with the more efficient extraction, refining, and distribution under people's nationalization, is a national public transportation system. Aside from its many other advantages it could greatly reduce the use of gasoline which in the United States currently surpasses one-half of the world's consumption.

With the funds that would be readily available from the monopolies it would be quite feasible to expand domestic production, not only by ending the contrived shortages but by rapidly expanding into the untapped areas of oil shale, coal gasification, and offshore exploration and development, to name only a few.

The current program to conserve energy through such measures as turning down the thermostats is essentially a fraud. Less than 30% of total U.S. energy consumption is used by the American people; the bulk is consumed by business, especially big business, of which an important portion is wasted by the military and in airplane junkets, yacht trips and other pleasures of the very rich.

Also high on the list of a people's agenda is expansion of trade with the USSR. The Soviet Union is the world's second largest producer of oil. The price of a barrel of Soviet crude is low and it is in high demand throughout the world.

What about international relations in oil? There is a sensible way to proceed in order to eliminate the simultaneous anarchy and monopoly operating in the purposeful chaos of the world oil market. Free competition in oil is a thing of the past. Most of the world's oil is now or will soon be produced by government corporations. The way toward creating a stable world petroleum trade is through long-term international commodity agreements. Such agreements were visualized as important means of creating stability in world commodity markets and were encouraged by the United Nations and affiliated bodies after World War II. There were limited, partial agreements in sugar, coffee, wheat, and some other commodities. But they ultimately collapsed, primarily because the United States and other imperialist countries tried to use them to perpetuate a one-sided relationship in which all the advantages flowed to them while the developing countries producing these commodities suffered from restricted markets and unreasonably low prices.

The oil lobby speaks about the "Arab embargo." Kissinger speaks about "blackmail." Until a few months ago the United States had a partial embargo on all foreign oil. It was lifted only because of the oil emergency. Moreover, the United States and its imperialist allies tried to keep the Soviet Union and other socialist countries out of these agreements which they used as weapons of economic warfare against them. Trade and exchange of commodities on a basis of equality were made impossible by the cold war policies of embargo tariffs and credit restrictions.

Apparently this is what Kissinger and Nixon have in mind now. They want to restore the old relationship through international agreements dominated by the imperialist countries, excluding the socialist countries, and in which the producing countries enter as weaker members carrying out the dictates of the United States and its allies.

But a proper international commodity agreement for petroleum would be to the mutual advantage of all countries, industrialized and developing, socialist and capitalist. Any other approach will fail.

Its components would include:

1. Long-term supply agreements by which each producing country would agree to supply specified quantities of oil to importers of oil.

2. Long-term commitments from the purchasing countries to deliver capital equipment and other goods required by the producing countries under suitable credit terms.

3. Stable prices agreed to for both the petroleum and the products required by producing countries, to be adjusted in an agreed manner for currency devaluations and changes in the general level of world prices.

4. An end to all lawsuits, ship seizures, freezing of balances, economic embargoes by imperialist countries to enforce compensation for nationalized properties in developing countries. All participants to accept without reservations the United Nations resolution of December 1973 on Permanent Sovereignty over Natural Resources.

A logical counterpart of such an agreement would be expanded, multilateral international cooperation in development of new forms of energy, in protection of the environment, and related questions.

Economic agreements of this sort would not guarantee economic stability in the capitalist countries. They could not overcome the internal contradictions within each, nor the consequences of uneven development among them because of the more rapid growth of some. Nor would such agreements end economic rivalry between individual capitalist countries and between the socialist world and the capitalist world. But they would contribute significantly to an improved supply of energy in all countries, to stabilizing living costs in capitalist countries, to speeding the industrialization of the developing countries. Such a commodity agreement in oil would go far to eliminate one of the main objects of imperialist war and wars for the national suppression of peoples. It would advance the cause of detente, of peaceful coexistence between capitalist and socialist countries, and create more favorable conditions for the struggle for peace and disarmament in the United States.

An Idea Whose Time Has Come

The idea of nationalizing the oil industry is not new. Trade unions passed resolutions and promoted the idea at the turn of the century. The Communist Party, USA, has advocated nationalization throughout its fifty-five years.

What is new is that nationalization is now becoming recognized as an absolute necessity. Millions now see it as the only way out of the energy crisis.

Some fifteen years ago the radical-left writer Harvey O'Connor wrote: "The truth is that there can be no public control of the oil corporations. They are too big. As in the case of other industries, railroads and utilities, when efforts are made to regulate them they wind up regulating the regulators. The only answer is public ownership and democratic management—the proposals socialists have been advocating for nearly a century."

Some months ago, after attending the U.S. Senate hearings on the oil industry, Clay Steinman and Robert Entman wrote in the *Nation*:

> Yet within the conventional free market parameters, there seems no way to cure this situation. The industry is so powerful and vast that we may have come to the point where *no reforms are adequate*, where without deep, fundamental

changes in our economic system, there is essentially nothing to do but allow the private sector greater power over public policy. One can argue that exactly that has been done; the oil companies have been allowed to decide how much fuel they will supply us and at what cost—and with little regard to rational current usage or prudent future planning.

Thus the independent power that seems to have accrued in America to capital concentration and technological expertise may have rendered independent, regulatory government a civics-book illusion. Congress holds hearings, but otherwise it seems ineffectual in the energy area. The Administration is closely allied with oil interests, and no previous administration could accurately be called anti-oil. Judicial remedies seem unlikely.

Perhaps, then, the oil executives and apologists are right when they say that the only way to get more oil for America under the present system is to pay the oil companies such an outrageously large ransom that they will respond by doing what they can to give the country all the fuel it wants. Although it was only hinted during the hearings, this adds up to a compelling case for nationalizing the energy conglomerates, and putting the nation's energy resources and its international dealings for fuel under the rule of rational policy. Reformist regulation, as the oil executives point out, most likely would only further complicate an already massive, intricate industry.*

Nationalization under democratic people's control can result in real gains for the American people. The monopolies know this all too well. That is why they have fought tooth and nail.

A glimpse of what nationalization could mean can be seen in the workings of the Tennessee Valley Authority, a government-owned operation created at the time of Roosevelt's New Deal. Although a far cry from what it was in the old days, the electricity rates charged by the TVA today to resident customers of 1.3

* Clay Steinman and Robert Entman, "The Sovereign State of Oil," *Nation* 26 January 1974.

cents per kilowatt hour are still one-half the average for the electricity industry as a whole.

The same is true of municipal government electric power production compared with the private utilities—rates about one-half those charged by these utilities that monopolize 87% of the U.S. electricity demand.

But the rates alone tell only a small part of the story. The TVA, for example, engaged in a whole range of activities that the private utilities would not touch with a ten-foot pole. It sponsored demonstration farming projects with modern agricultural techniques that revolutionized farming throughout the whole area. It sponsored a wide variety of mineral exploration and development projects, and educational activities such as libraries and extensive technical research services. It provided construction jobs for hundreds of thousands of victims of U.S. capitalism's greatest economic catastrophe—the Great Depression of the 1930s. More than 6,000 new industries have been created with jobs for 800,000 people in the area served by TVA.

On-the-job accidents have been kept to a minimum. The TVA has won first place in the annual rankings of the National Safety Council on accident frequency or severity on large construction projects.

And the dams the TVA built were constructed in record time, producing 1.5 times the per capita national average of electricity in a region that ten years earlier, produced only one-half of the national per capita.

In 1933, when the TVA started, only two major dams were on the Tennessee River and eight of its tributaries. The TVA built twenty-three new ones, including minor structures. Today fifty dams make up the TVA River Control System which extends 650 miles from Knoxville to the Ohio River.

When the TVA began its work in 1933 there was little provision for outdoor public recreation in the whole valley. There was no state park system in Tennessee, no county had provided for park areas; neither Tennessee nor Alabama had a state department of conservation. In 1934 the TVA developed several demonstration parks on land that lay near its reservoirs. Thousands of people in the valley visited these demonstration parks and approved the idea. They were ready to support efforts to further park developments.

Today the TVA reservoirs have an 11,000-mile shoreline, along which are 100 public parks and recreation facilities with an investment of nearly $400 million.

Although owned by the federal government the TVA was a local effort, and the same is true for municipal electric power. How much greater is the potential for a nationwide effort under people's power—a National Energy Council!

The ideologists for the oil monopolies work hard to frighten people away from nationalization. "It will bring chaos," they say. Their arguments are weakened by the *very real chaos* of the energy crisis. Life has prepared the soil for nationalization. The monopoly structure itself is part of the preparation.

In a different period, in a different country, Lenin, the leader of the first socialist revolution, stated the case very strongly:

> Take the oil business. It was already to a vast extent "socialized" by the earlier development of capitalism. Just a couple of oil barons wield millions and hundreds of millions of rubles, clipping coupons and raking in fabulous profits from a "business" which is *already* actually, technically and socially organized on a national scale and is *already* being conducted by hundreds and thousands of employees, engineers, etc. Nationalization of the oil industry could be effected *at once* by, and is imperative for, a

revolutionary-democratic state, especially when the latter suffers from an acute crisis and when it is essential to economize national labour and to increase the output of fuel at all costs.*

It was a new idea. It has been tested. And it works.

What would be some of the main benefits to the people from nationalized energy industries, run by a genuinely democratic National Energy Council?

1. *Inexpensive Energy.* Energy products would be sold to the people at cost, without profits, and without special excise and sales taxes.

2. A labor policy that meets the most advanced trade union standards. Five dollars per hour *minimum wage* at all levels of the industry would be paid. No refinery or tank farm could operate without meeting safety and health standards prescribed by the unions. The nationalized energy industries would be required to hire Blacks, Chicanos, Puerto Ricans, Native Americans, Asians *at all levels,* in proportion to their numbers in the population, and within a brief period of time.

3. Proper financial compensation for the native peoples of Alaska, and retroactive compensation to the Indian peoples of the contiguous forty-eight states for oil lands taken from them. Protection of the hunting and fishing lands of the Alaskan peoples, and guaranteed employment to all of them wishing jobs.

4. Adherence to environmental standards as a primary requirement in development of the energy industries.

Nationalization will not come easily. Monopoly corporations will not give up their privileged positions until they are forced to. Resolutions for

* V. I. Lenin, *Collected Works* (Moscow: Progress Publishers), vol 42, p. 38.

nationalization are an important first step. But they will become a reality only because of an organized people's movement.

There is a growing anticorporate sentiment in the country. But sentiment, no matter how deep, also will not do it. The sentiment must be organized into a broad anticorporate coalition, which must include all sections of the population that in one way or another are victims of monopoly oppression.

This broad coalition must be both grass roots and national in character. The sentiment is ripe. What is needed are initiatives and organization.

14 • Energy and Socialism

Nationalization of the energy complex would be an important step in the right direction. Yet it would not solve all of the problems because the publicly owned sector would continue to operate surrounded by capitalism. The Rockefellers, Mellons, and Chase Manhattan Bank would still be in business. The struggle against the monopoly corporations would have to continue.

Nationalization would be a meaningful step at a moment when the majority of people are not yet convinced of the next step that would make social properties of all industries, all banks, all transportation, the mass media, the energy complex, and lay the foundation for a new social and economic system. The corporate monopoly system, based on privately owned industries, must be replaced by a system based on the premise of a socially owned and operated industry. With this change the dog-eat-dog sense of values of an economic system based on legalized extortion, a system that ignores all human values in the quest for corporate profits would be replaced with a system propelled and motivated by the sole desire to do the best for the most, a system in which abundance is abundance for all.

The evils of capitalism make socialism an historical necessity. The struggle for social progress makes the elimination of capitalism an historical necessity.

Socialism eliminates the basic flaw, the main contradiction of capitalism, between the privately owned but socially operated means of production. Socialism plans production and distribution with the single purpose of satisfying the needs of the people.

The elimination of capitalism will for the first time permit human society to advance without the ball and chain of corporate greed. The basic problems that have surfaced with the energy crisis cannot be solved by reforming capitalism. They are only additional arguments for the necessity of socialism.

Today there are two social and economic systems motivated by opposite sets of values. While taking into account that capitalism has been in the process of development for more than two hundred years and socialism appeared on the world scene only some fifty-five years ago, it is still very useful to make general comparisons.

It is a logical question to ask: "Why is it that there are no energy crises, no unemployment, no runaway inflation, no economic depressions or crises in the socialist countries?" At a moment when the capitalist world is in the throes of continuous instability, the socialist countries are in the process of growth in an atmosphere of stability. The corporate ideological complex puts out such a barrage of anticommunist propaganda that sometimes it is difficult for the public to separate facts from the falsehoods.

The socialist countries have not solved all social and economic problems. Very often the past is a drag on progress. But they have resolved the fundamental questions of economic and social structure, and have eliminated the basic contradictions that plague capitalist societies, the main inequalities and injustices. They have eliminated exploitation, which is the fundamental basis for inequalities and injustices. They have the structure upon which the future human society is being built. Capitalism provides no such structure. Increasingly, capitalism becomes an obstacle to the solution of problems facing humanity. Because the Soviet Union has been building the new socialist society longer than any other country it is instructive to take a closer look at how they are solving the problems of energy.

The Socialist Solution

The Soviet Union, as the first socialist country, faced an acute energy shortage from its founding. The Civil War resulted in the flooding of coal mines and wrecking of oil wells—such is the mentality and morals of capitalism. The development of the electric power industry was still in its infancy. Large numbers of engineers and technical specialists emigrated. The previously normal sources of equipment and technique in Western Europe and the United States were cut off, embargoed.

In World War II its energy industries again suffered severe damage at the hands of the Nazi invaders, while normal technical progress and growth was interrupted during the ten years of war and reconstruction.

Yet today, the Soviet Union is close to parity with the United States in total production of energy. It leads the world in production of coal, and is a very close second in production of oil and natural gas. Soviet geologists, geophysicists, and drilling teams in

Siberia have developed some of the world's greatest reserves of oil and natural gas; and Soviet workers and engineers have accomplished real miracles of production in some of the world's most difficult terrain, extracting and bringing to market huge quantities of this material in a relatively few years.

The Soviet scientists and engineers overcame the economic and scientific technical embargo of the United States to develop first-rate drilling, transporting, and refining techniques, including sophisticated advanced refining methods and large-scale petrochemical production.

Besides supplying the energy needs of its own rapidly advancing economy the Soviet Union has supplied the bulk of the oil and gas needs of the East European socialist countries and Cuba, as well as substantial quantities of the needs of Western Europe and a number of developing countries, such as India. After the Chinese Revolution the Soviet Union furnished and built a self-sustaining oil industry in People's China.

The Soviet Union was the *only* country to develop atomic energy independently. The United States got its head start thanks to the services of the outstanding scientists of Western Europe, refugees from fascism, as well as from being spared any of the damages of World War II. By the outstanding efforts of its own scientific personnel and workers the Soviet Union soon ended the atomic monopoly and foiled the atomic political extortion of U.S. imperialism—and its openly announced plans to invade and conquer the Soviet Union with the aid of atomic weapons. By 1940 the Soviet Union had mastered atomic energy, which with its other economic accomplishments, secured it against the savage threats of John Foster Dulles and all the vultures of U.S. imperialism plotting its destruction.

Moreover, the Soviet Union has paralleled the United States in development of atomic energy for peaceful uses, and is now in the process of bringing into production an increasing number of large atomic energy power plants apparently technically more reliable and considerably safer than their U.S. counterparts. Up to now the Soviet Union has led the world in the research work leading to production of thermonuclear power, which promises to provide an unlimited source of cheap energy in the twenty-first century.

The United States has spent a cumulative sum of some $400 million on fusion research. The Soviet Union has spent about $800 million.

Fifteen percent of all fusion scientists in the world are in the United States—many of them are unemployed. Thirty-five percent of all fusion scientists of the world are in the Soviet Union—all fully employed.

Of course the Soviet Union is aided in its energy development by a wide expanse of territory and a large population. But this alone certainly does not account for its ability to far surpass most imperialist countries including countries that were well advanced in science and technique at the time of the Russian Revolution.

The Soviet Union is virtually unaffected by the energy crisis, by the wild inflation in fuel prices sweeping the capitalist world. That applies also to the socialist countries getting their supplies from the Soviet Union, and there is no multiplication of prices charged consumers of gasoline, oil, and coal.

The decisive fact here is the advantage of a socialist, planned economy. The founders of the Soviet state knew the key importance of the energy industries as succinctly formulated in Lenin's famous phrase: "*Communism is Soviet power plus the electrification of the whole country.*" In 1920, as the Civil War was drawing to a close, Lenin sponsored the first plan for the electrification of Russia, the Goelro Plan:

> We have already drawn up a preliminary plan for the electrification of the country; two hundred of our best scientific and technical men have worked on it. We have a plan which gives us estimates of materials and finances covering a period of years, not less than a decade. . . .What we must now try is to convert every electric power station we build into a stronghold of enlightenment to be used to make the masses electricity-conscious, so to speak. . . .To carry out the electrification plan we need a period of ten or twenty years to effect the changes that will preclude any return to capitalism. This will be an example of rapid social development without precedent anywhere in the world. The plan must be carried out at all costs, and its deadline brought nearer.*

The plan was carried out ahead of time, and today the Soviet Union is second only to the United States in production of electricity.

The Soviet Union is forging ahead in the development of the magnetohydrodynamic method (MHD) of electic power production—relatively unheard of in the United States—which converts heat directly into electrical energy. Two high-capacity installations have already been built and specialists believe that MHD installations are a promising source of energy for the near future.

The advantage of MHD electic power stations is that they can reach an efficiency of up to 65% compared to 40% to 42% for normal thermal stations according to Soviet estimates.

Soviet production of basic commodities was cut sharply during the Nazi invasion of World War II. In 1945 production of coal amounted to 149 million tons, of oil 19 million tons, and of steel 12 million

* V. I. Lenin, *Collected Works* (Moscow, Progress Publishers, 1966), vol. 31, pp. 516-517.

tons. The Soviet government set the goal of raising these figures to 500 million tons of coal, 60 million tons of oil, and 60 million tons of steel by 1960, or as soon thereafter as possible.* Clearly they also saw the key role of energy in the industrial might of a country. When these levels were reached, they said, the Soviet Union would be safe against all contingencies—that is, against a military assault by hostile imperialist powers.

That inveterate enemy of the Soviet Union, *New York Times* Sovietologist Harry Schwartz, characterized these as "ambitious goals," and admitted that in striving to reach them "stupendous achievements were realized" in the five years immediately after the war.**

The actual amounts realized by 1960 were: for coal—510 million tons, steel—65 million tons, and oil—148 million tons, or 2½ times the goal!

The Soviet Union is second to none in power technology, in the range and sophistication of its electric power producing and transmitting equipment, and in the assistance it provides to other socialist countries and to developing countries in creating their own electric power industries.

Soviet energy planning was based decisively on using its own natural resources. Of course the oil of Iraq and Iran is easier to get at than the oil of Siberia, but the Soviet Union, as a socialist country, could not consider plundering the resources of other countries. It has no private corporations, so there is no interest in seizing these resources for private profit. But, one may say, the Soviet Union didn't have the opportunity to seize Iraqi and Iranian oil. But it did have the opportunity to seize Romanian and Hungarian oil and

* Soviet production figures are in metric tons, about 10% larger than the net tons in U.S. production figures.

** Harry Schwartz, *The Red Phoenix*, 1961, pp. 112-113.

Polish natural gas. It didn't touch a drop. Instead it cooperated with these countries in the development of their resources. It gave decisive assistance to People's China in developing the famous Taching oil fields and didn't take a drop. These are policies that flow naturally from the inherent character of socialism.

In recent years the Soviet Union has purchased oil and natural gas from its southern neighbors and has long-term contracts with Iran for natural gas deliveries. But these are on a basis of equality, with Iran retaining ownership of its gas resources, and the Soviet Union, in exchange, supplying Iran's first steel mill and other basic industrial installations.

The Soviet Union, therefore, enters the present crisis as the ally of the oil producing countries and is increasingly recognized as such. It is in no way damaged by the struggle of these countries against the multinational oil cartel.

Development of energy supplies in the Soviet Union and other socialist countries is integrated into an overall economic plan.Thus, priority had to be given to energy consumption to build basic industry, commercial transport, agriculture, as well as for fueling a defense establishment that could match the most powerful and aggressive imperialist powers, providing the basic energy needs of other socialist countries without significant natural resources, and for trading purposes with capitalist countries able to supply needed equipment.

For many years the U.S. media has mocked the Soviet people for their lack of private automobiles. Yes, the private automobiles had to come later because without satisfying the priority needs the Soviet Union would not be able to support a motorized private economy. But private automobile production is now being expanded rapidly as the reserves of energy and other materials make it possible.

Meanwhile the Soviet people have benefited from the best and cheapest public transportation system in the world, and now will have added to it an increasing supply of the more flexible private vehicles. Private motorization of the economy will be planned and subject to sensible restraints; it will not lead to the excesses of numbers of cars, gas shortages, traffic jams, and pollution characteristic of the capitalist countries. Because of its socialist character the main emphasis will continue to be on improving mass transit systems.

The present energy crisis brings to the forefront this aspect of our economic relations with the Soviet Union. While giving priority to its own needs and those of other socialist countries, the Soviet Union provides in its planning for significant exports of oil and gas to capitalist countries in exchange for equipment and materials it needs. The Soviet policy is to trade with the capitalist countries at world market prices, and to seek agreements for the stabilization of those prices. Vast, multibillion-dollar deals involving huge quantities of industrial products have been concluded with West Germany, Austria, Italy, and France.

The Soviet Union sells small quantities of oil to Japan. But for many years Japan has refrained from finally settling its endless negotiations for Soviet oil and gas. A powerful influence here has been the pressure of U.S. imperialism on Japan not to conclude such a deal, and in effect to let U.S. companies continue to have a virtual monopoly of the Japanese market.

For two years now the Soviet Union has been negotiating with consortiums of six U.S. companies and a Japanese group for two projects—one all U.S. and the other U.S.-Japanese—involving shipments of unprecedented quantities of liquefied natural gas to the east and west coasts of the United States. After considerable hesitation, the Nixon administration gave its approval to these projects, in principle, but they have been subject to a bitter opposition campaign from the chronic anti-Soviet elements. The Pentagon, the ultraright, the Zionist lobby, are today the principal sources of this opposition.

A Leader of the Anti-Detente Pack

Their most persistent and pseudoliberal mouthpiece has been the *New York Times*. This newspaper, which today as in the years immediately after the Russian Revolution leads the pack in anti-Soviet incitation and slander, features two arguments. It says first that Soviet natural gas will be too expensive, even though there have been no published figures on the range of prices under negotiation.

Naturally, the *New York Times* would prefer dealing with a weak, neocolonial country that would give up its natural resources to private corporations for next to nothing. But recent events show that these days are gone forever. Whatever the prices discussed earlier it seems clear that the Soviet gas will be no dearer and perhaps cheaper than the prices U.S. monopolies are seeking and all too likely will pay in the near future. It is interesting that several years ago the Japanese canceled prolonged negotiations for Sakhalin natural gas over minor differences on prices, according to Japanese accounts. The *Times'* complaints about Soviet prices for natural gas look foolish under existing conditions and it no longer emphasizes that line.

The *Times* gives more play to its second argument: the USSR is an unreliable trading partner, that it is apt to hold up supplies to the United States for political reasons. Supporters of the deal point out that the proposed Soviet supplies will not be more than 5% of total U.S. consumption and an interruption for any reason would not be crippling.

If there is hesitation on political grounds it surely is more serious on the Soviet side. The United States time and again has outrageously broken contracts with the USSR and other socialist countries for political reasons. It has engaged, and continues to engage, in the most outrageous economic discrimination against these countries; and still uses economic warfare freely against countries taking an anti-imperialist position.

The offer by the Soviet Union of long-term supply contracts signifies its estimate that the prospects of detente and an atmosphere of peaceful coexistence are sufficiently good to justify the risk of losses due to possible future disruption on the part of the United States.

The opponents of such trade may be in cahoots with oil and gas company elements in this country anxious to avoid new sources of supply that would hamper their attempts to manipulate shortages. However, the main motive is undoubtedly political—the desire to halt and reverse the process of detente.

The United States will lose more than gas if it turns down the Soviet LNG deal. Nicholas J. Coolidge, a vice president of Kidder Peabody & Co., explained why he and other Wall Street investment bankers are so anxious to conclude the deal. Explaining that it would be best financed by bonds floated through the Wall Street houses which would get a high underwriting charge, he said, "The USSR would be able to enjoy interest carried on prime U.S. government triple A issues,"—naturally, because the bankers know that the Soviet government is at least as reliable as the U.S. government. Coolidge gave his estimates for one of those giant developments, the North Star Project, which is sponsored by Tenneco, Texas Eastern Transmission, and Brown and Root. They plan to bring about 2 billion cubic feet of gas per day 1,500 miles from the Urengoi area of western Siberia to

Murmansk, where it would be liquefied and then carried in twenty tankers to the U.S. east coast. The overall cost may exceed $6 billion.*

Assuming, with Coolidge that half of this money would be spent in the United States, there would be at least 100,000 man-years of work created—to build the vessels, the equipment for piping, liquefying, and storing the gas that would be delivered to the Soviet Union and used in this country, and in technical cooperation with Soviet specialists engaged in the project. In addition, there would be continued large-scale employment transporting and deliquefying the gas, in the follow-on deals with the Soviet Union as it received payment for delivery of gas, and as it required replacements, and even more as the successful results of this project led to more and even bigger cooperative projects.

United States' stalling on the project does not endanger the development and export of Urengoi gas. It simply endangers any U.S. part in it. Theodore Shabad reports in the *New York Times:*

> The Soviet Union, which has been looking to American investment in its Siberian oil and gas resources, has apparently not been deterred by the prospect of cutoff of large credits from the United States and has been moving forward on several fronts through its own efforts.**

As for the second giant project, for the gas of Yakutia, recently the Japanese have been indicating an inclination to go ahead with it, with or without the United States. Whether they will do so or yield to the pressure of Washington remains to be seen.

The main losers of a failure to participate will be the capitalist countries and their people. As far as the American people are concerned it is in our interest, both economic and political, to combat the line of the

* *Money Manager,* 10 December 1973.
** 8 January 1974.

New York Times and to actively support the projected energy arrangements with the Soviet Union, along with the complete execution of the 1972 and 1973 economic agreements between the two countries, now being blocked by the anti-Sovieteers in Congress and their supporters.

The Nixon administration cannot be relied upon to carry on a fight against these elements any more than it has in the past. The Nixon-Kissinger line of excluding the socialist countries from all efforts to develop energy cooperation indicates a continued effort to use energy as a weapon of economic warfare against the socialist world and, even more, to assist the oil monopolies to reassert and strengthen their grip on the capitalist world's oil. This would be impossible under conditions of cooperation involving the socialist countries as well as the industrialized capitalist countries and the developing countries.

15 • Marxism—a "Computer" for Victory

To get at the basic truth, the real causes of the energy crisis, it is necessary first to absolve nature from any responsibility. Nature is not the culprit and it is not bankrupt. Nature, matter in motion, is nothing but energy. For example, the sun has been showering the earth with its life-giving energy for some 5 billion years. And from all indications it will continue in the energy business for at least another 10 billion years. The earth will continue to bask in the energy rays of the sun unless we so pollute the atmosphere that there will be an energy lockout. That is one of the challenges human society will have to accept.

Up to the present moment human society has used the energy sources that have been readily available. In a sense all that has been necessary is an axe for wood, a shovel for coal, and a pump for oil. It is this readily available energy that in some areas is now beginning to run out, especially in countries such as Britain, France, and West Germany.

Obviously, the rate of consumption of energy will continue to grow. The present rate of growth worldwide is 5.5% annually. As the underdeveloped countries pick up steam, building their industries and improving the overall quality of life, the rate of energy consumption will increase (in the United States the amount of energy used has doubled every fifteen years). For this rising rate it is clear that new sources will have to be found. Or, to put it more accurately, the new sources are known, but what is needed is the new technology to turn these sources into usable power. Nature has the potential for an inexhaustible supply of energy.

The new technology will not only have to provide the extraction of energy, it must be able to cope with the new level of environmental consequences from its use, extraction, and consumption.

To get at these new sources of energy requires breakthroughs in science and technology not possible in the scattered, private , corporate laboratories. It will come about only as a result of a planned, synchronized assault on a national or global scale. It will not happen if it is tied to a price tag or private corporate profit margins. It will not happen as long as the science community is militarily oriented.

The energy crisis cannot be separated from the *general crisis* of science and technology under capitalism, particularly under U.S. state monopoly capitalism. This crisis is manifest in the falling off of real dollars for basic research, in the declining number of students in science, in the growing unemployment among scientists. The historical reason for this is the basic contradiction between the social and increasingly scientific-technological means of production, and the private, corporate expropriation of the end product.

The anarchy of capitalist production is a built-in obstacle to a scientific-technological level that needs planning and centralized control. Corporations have for years built into their products a planned obsolescence, and now the crisis is exposing the built-in, inherent obsolescence of capitalism as a system increasingly in contradiction with the peoples' expanding energy needs.

For example, under these conditions the environment cannot be protected from negative effects by patchwork treatment. Protection cannot be an afterthought. To be effective it must become a built-in, planned feature of the production of energy itself. Environmental protection will have to become an integral feature of every branch of science.

Such problems cannot be left to the discretion of a few profit hungry corporate executives.

Frances Sargent, the governor of Massachusetts, reflected, ". . . one cannot and should not expect the energy companies to be the guardians of the public interests—they are simply profit maximizing institutions operating in a quasi-monopolistic market and they are going to act as such." And not to close the doors to the election campaign contributors he adds the phrase: "Nor do I mean to imply that the gas and oil companies are out to exploit the consumer."*

The profit-maximizing corporations and the Sargent-type politicians are obstacles to the solution of the longer range energy problems.

The obstacle to the path of unlimited energy is the present social structure. When asked why his corporation had not drilled for oil in the United States, a leading spokesman of one of the top ten oil corpora-

* *The National Energy Problem*, Academy of Political Sciences, December 1973.

tions replied: "To some people I know that is going to sound evil. But we have a business proposition here. It is hardheaded business economics." That is the obstacle, "business economics" which is never concerned with human welfare or with social or national interests.

The Gulf Oil Corporation gives out U.S. flags as decals and bumper stickers; it promotes bumper sticker patriotism. When one of its top executives was asked why Gulf has not drilled for oil, or why it has not built refineries in the United States, he said: "The oil companies spent the money overseas where the investments were good." And he continued, "I think it is quite a logical thing to have done." In business economics the human need for products is something to be used for making profits; the "logical thing" is private profits, not human concerns.

It is the socio-economic structure based on the evil of business economics that is the basic cause of the present energy crisis and the obstacle to its solution; and as the production of energy becomes more difficult it will become an increasingly bigger obstacle.

Solar and atomic energy and the business economics of capitalism are incompatible, an insoluble contradiction. The frustration and the dilemma that it presents were expressed well by Prof. David C. White of MIT:

> And we know that decisions based on cost benefit analysis at current economic interest rates, though they may yield conclusions valid for profit-making industries that necessarily have a restricted domain of concern, are not necessarily best for society's long term interest, nor capable of dealing with future uncertainties. So we conclude that major institutional revisions to allow proper interactions between the economic marketplace, research and development and the long-term interests of society, each with a different constituency having different goals and different dynamic response characteristics further complicated by

high degrees of uncertainty, are a first requirement for dealing with the energy crisis.*

The long-term interests of society can be served only by a social structure that is motivated by the long-term and the short-term interests of society —socialism. A capitalist structure based on and motivated by narrow private profit interests is inherently an obstacle to a solution of the energy problems.

The Point of No Return

The energy crisis is stimulating provocative thoughts about where human society is at and what are the options before it. The sense of values and priorities, taken for granted and passively accepted by generations, are now seriously questioned.

For fifteen months C. Jackson Grayson was the chairman of Nixon's Economic Stabilization Program, providing him with an unusual opportunity for an insider's view of the capitalist establishment. His conclusions are rather pessimistic. In *U.S. News and World Report,* he asks the question: "Has the erosion of the free market economy reached a point of no return?" He answers his own question: "At some point—and I predict that at the present rate this point may be reached in about fifteen-twenty years—the essential characteristics of a competitive private enterprise system will no longer make up the economic engine that drives our system."**

Mr. Grayson is correct. There is "erosion" and it is a feature of a law-governed historic process. It is also correct that each stage of its erosion is "a point of no return." There is no way capitalism can be restored to its premonopoly days. There are momentary crises

* Energy—Environment—Economic Triangle, *Technology Review,* December 1973, p. 11.

** *U.S. News and World Report,* 14 January 1974.

that come and go, but the "erosion" Mr. Grayson senses is a general crisis, an irreversible process of decay.

Ten years ago the executives of the oil corporations and banks fed data of the policies they were going to follow into their computers. The computer projections were "an energy crisis." Production and consumption of energy were on a collision course.

But over one hundred years ago Karl Marx said capitalism would give rise to just such corporate monopolies which at a certain stage would have the power and, in their drive for "maximizing" private profits, would dictate policies that would lead to crises—such as the energy crisis. Marx's projections obviously were of much greater significance than the corporate computer's projections. The Exxons, General Motors, and the Chase Manhattan banks were clearly outlined in Marx's projections. Marx foresaw the "erosion" and "the point of no return."

All of the Exxons and the Chase Manhattan banks of big business of today are the recipients of the inherited characteristics of capitalism. They are creatures of the inner laws of capitalist development. It was these inherited characteristics and the operation of these objective laws that Marx and his co-worker Frederick Engels discovered. On the basis of these laws it was possible for them to foresee and project the direction of capitalist development.

The results of these inherent laws of capitalist development are like a broken-down car. It is not just a bad tire or a faulty horn. The engine, transmission, and other basic systems have fallen apart. The ruin of one results in the destruction of others. These sputtering rejects interact with each other causing a total wreck of an automobile. By spending a lot of money you can make it run for short distances, but you are better off without it.

The mad drive to maximize profits, vast monopolization, monopoly takeover of the government and development of a police state structure, crises, the multinational mutilation of nations and peoples abroad—all of these interact with each other producing a huge lemon of a social system. To this should be added the continuous inflation, the racist monopoly superprofits, and the bloated spider of finance capital—the Morgans, Rockefellers, Mellons, etc.—perched on top and pulling the financial strings.

With capitalism we are dealing with an economic system that has served its historical purpose. It has outlived its usefulness. It has given birth to a class of corporate thieves who do anything and everything to perpetuate a setup that makes it possible for them to continue their gigantic rip-offs.

Marx's "computer" has developed into the social science now known as Marxism-Leninism. Its data banks include the very best of all accumulated human knowledge. It has assimilated all of the world experience of struggle for human progress. Its data banks show that what is needed is a change in the corporate life-styles—a change from the life-style of the leech to the life-style of a productive worker. What is needed is a change of system—from an old and eroding capitalism to a new and vigorous, full of promise socialism.

APPENDIX • A

The Global Shell Game

WESTERN HEMISPHERE

USA—Shell Oil Company, Asiatic Petroleum; **Canada**—Shell Canada, Canadian Fuel Marketers; **Caribbean**— Conch International Methane (Bahamas), Shell Bahamas, Shell Bermuda, Shell Antilles & Guianas, Refineria Dominicana, Shell (W. I.) ((Dominican Republic, Haiti, Jamaica, & Panama)), Shell des Antilles et de la Guyane Francais, Antillaise d'Entreposage (Martinique), Raffinerie des Antilles (Martinique), Shell Curacao, Shell Nederlandse Antillen, Pipelines of Puerto Rico, Shell (Puerto Rico), Shell & Commonwealth Chemicals (Puerto Rica), Shell Trinidad; **Central America**— Shell British Honduras, Shell Costa Rica, Refineria Acajutla (El Salvador), Shell El Salvador, Refineria Petrolera de Guatemala-California, Guatemalteca Shell, Shell de Exploracion Honduras, Distribuidora Shell de Mexico, Shell de Exploracion Nicaragua, Shell Nicaragua. **South America**— Shell Argentina de Petroleo, "Estrella Maritima" (Argentina), Shell Brasil, Shell Chile, Shell Condor (Colombia), Shell Colombia, Ecuatoriana de Lubricantes, Shellgas (Ecuador), Shell de Recherches et d'Exploitation de Guyane (Fr. Guiana), Guyana Shell, Shell Paraguay, Shell del Peru, Shell Suriname Verkoop, Shell Uraguay, Shell Sur del Lago (Venezuela), Shell de Venezuela, Shell Quimica de Venezuela.

EUROPE

Austria— Rohoel-Gewinnungs, Adria-Wien Pipeline, Transalpine Oelleitung in Oesterreich, Shell Austria;

Belgium— Belgian Shell, Distragaz, Bayer-Shell Isocyanates; **Denmark**— Shell Denmark, Dansk Shell; **Finland**— oy Shell ab; **France**— Shell Francaise: Maritime Shell, Raffinage Shell Berre, Petroles Shell Berre; Francaise de Stockage 12 G, Geologique, Shell-Gascogne, Petroles de Sud-Est Parisien, Pipeline du Jura, Pipe-Line Sud-Europeen, Rhenne de Raffinage, Produits de l'Air, Utilisation Rationele de Gaz, Shell Chimie, Chimique de la Mediterranee; **Gibralter**— Shell Gibralter; **Greece**— Shell (Hellas); **Irish Republic**—Irish Refining, Irish Shell and BP; **Italy**— Italiana per l'oleodotto, Transalpino, Shell Italiana, Sub-Sea Oil Services; **Luxembourg**— Shell Luxembourgeoise; **Malta**— Shell (Malta); **Netherlands**— Shell Delstoffen, Nederlandse Aardolie, Internationale, Riviertankscheepvaart, Rotterdam-Rijn Pijpleiding, Shell Tankers NY, Shell Nederland Raffinaderij, Shell Nederland Verkoop, Nederlandse Gasunie, Internationale Gas Transport, Rotterdamse Polyolefinen, Shell Nederland Chemie, Verenigde, Kunstmestfabrieken, Wavin, Shell Research, N.Y., Billiton; **Norway**— Norske Shell; **Portugal**— Shell Portuguesa; **Spain**— Shell Spanje, Espanola Shell, Industrias Quimicas Asociadas; **Sweden**— Koppartrans, Svenska Shell; **Switzerland**— Oleoduc du Jura, Neuchatelois, Raffinerie de Cressier, Shell (Switzerland); **United Kingdom**— Shell U.K. Exploration & Production, Shell International Marine, Shell Tankers (U.K.), United Kingdom Oil Pipelines, Shell U.K., I.B.E. Ltd., Shell Composites, Shell-Mex and B.P., Lubricants Producers, Shell Chemicals U.K., Shellstar, Associated Octel, Shell Research Ltd., **West Germany**— Gewerkschaft Brigitta, Gewerkschaft Elwerath, Deutsche Shell Tanker, Deutsche Transalpine, Oelleitung, Rhein-Donau Oelleitung, Rhein-Main, Rohrleitungstransport, Deutsche Shell, Gewerkschaft Deurag-Nerag, Ruhrgas, Thyssengas, Deutsche Shell Chemie, Rheinische Olefinwerke.

AFRICA

North Africa— Algerienna du Methane Liquide, Sirtica Shell (Libya), Shell Exploratie en Productie (Libya), Shell du Moroc, Shell de Tunisie; **West and Equatorial Africa**— Shell West Africa (Angola & Gambia), Shell Camerounaise, Shell de Cameroun, Shell de l'Afrique

Equatoriale, Shell Congo Kinshasa, Congolaise des Petroles Shell (Kinshasa), Shell Gabon, Equatoriale de Raffinage (Gabon), Shell Ghana, Shell Guinee, Shell Ivoirienne, Ivoirienne de Raffinage, Shell de l'Afrique, Occidentale, Shell-BP Nigeria, Nigerian Petroleum Refining, Shell Nigeria, Africaine de Raffinage (Senegal), Shell Senegal, Sierra Leone Petroleum Refining, Shell Sierra Leone; **East Africa—** Petroles des Djibouti, Shell Ethiopa, BP-Shell Kenya, East African Oil Refineries, Kenya Shell, Shell Chemical Eastern Africa, Maritime de Madagascar, Malgache de Raffinage (Malagasy Republic), Shell de Madagascar et des Comores, Shell Co. of the Islands (Mauritius and Seychelles), Shell Mozambique, Shell Red Sea (Hedjaz), Shell de la Reunion, Shell and BP (Sudan), Shell Sudan, Shell and BP Tanzania, Shell and BP Uganda, Shell and BP Zambia; **Central and Southern Africa—** Shell Botswana, Shell Lesotho, Shell (Malawi), Shell Chemical Central Africa, Central African Refineries (Rhodesia), Shell Rhodesia, Shell Eksplorasie Suid-Afrika, Shell and BP South African Refineries, Shell South Africa, Shell Chemical South Africa, Shell Eksplorasie, Suidwes-Afrika, Shell South West Africa, Shell Swaziland.

MIDDLE-EAST

Abu-Dhabi— Abu-Dhabi Petroleum; **Cyprus—** Cyprus Petroleum Refinery, Shell Cyprus; **Iran—** Iranian Oil Exploration and Producing, Iranian Oil and Refining, Naft Pars, Shell Chemical Iran; **Iraq—** Basrah Petroleum, Iraq Petroleum, Mosul Petroleum; **Kuwait—** Kuwait Shell Petroleum Development; **Lebanon—** Shell Lebanon; **Oman—** Oman Shell, Petroleum Development (Oman); **Qatar—** Qatar Petroleum, Shell Qatar, Shell Markets (Middle East); **Trucial States—** Shell Hydrocarbons, Shell Minerals; **Turkey—** Turkse Shell, Anadolu Tasfiyehanesi, Shell Turkey.

FAR EAST AND AUSTRALASIA

Brunei— Brunei Shell, Brunei LNG; **Cambodia—** Shell du Cambodge; **Ceylon—** Shell Ceylon; **Hong Kong—** Shell Hong Kong; **India—** Burmah-Shell Refineries, Burmah-Shell India, National Organic Chemical Industries;

Indonesia— Djawa Shell, Shell Indonesia; **Japan**— Nishi Nihon, Shell Sempaku, Seibu Sekiyu, Showa Sekiya, Showa Yokkaichi, Shell Sekiyu, Mitsubishi Yuka, Shell Kagaku; **South Korea**— Korea Shell, Kukdong Shell Oil, Kukdong Shell Petroleum; **Laos**— Shell du Laos; **Malaysia**— Sabah Shell, Sarawak Shell, Shell Refining (F.O.M.), Shell Malaysia, Shell Malaysia Trading, Shell Marketing Borneo; **Pakistan**— Pakistan Shell Oil, Pakistan Refinery, Pakistan Burmah Shell, Titas Gas, Burshane (Pakistan); **Philippines**— Shell Philippines, Shell Chemical (Philippines); **Singapore**— Shell Eastern Petroleum, Shell (Singapore); **Thailand**— Shell Thailand; **South Vietnam**— Shell Viet-Nam; **Australia**— Shell Development (Australia), W.A.G. Pipeline, Shell Refining (Australia), Shell Australia, Shell Chemical (Australia); **New Zealand**— BP Shell Aquitaine and Todd, BP Shell Todd (Canterbury) Services, Shell BP and Todd, Shell and BP Pipeline Services, New Zealand Refining, Shell Oil New Zealand; **Pacific Islands**— Shell (Pacific Islands), Shell des Iles Francaises du Pacifique; **Tonga**— Tonga Shell N.Y.

APPENDIX • B

A RECORD OF "JOINT VENTURES" OF OIL MONOPOLIES IN DIFFERENT PARTS OF THE WORLD.*

AFRICA

Throughout the continent, Gulf and Exxon have formed the United Petroleum Securities Corp. which in turn owns a controlling interest in a French corporation that refines and sells oil products in Europe and Africa.

In *Rhodesia*, Shell, British Petroleum (BP), Mobil, Texaco, and Standard Oil of California, hide under the corporate name of Central African Petroleum Refineries (PVT) Ltd.

In *Algeria*, Continental Oil and Shell operate a natural gas plant under the name of Conch International Methane Ltd. In *Libya*, Amerada Hess, Continental Oil, Marathon and Shell engage in mining, refining, and transport under the name of Oasis Oil Co. of Libya, Ltd. In *Kenya*, BP and Hess operate through the BP and Shell Petroleum Development Co. of Kenya. In *Nigeria*, these same two companies have teamed up under the name of Shell and BP Petroleum Development Co. of Nigeria.

In *South Africa* these same two companies operate a refinery through Shell & BP South African Petroleum Refineries. In the *Sudan*, they have formed Shell & BP (Sudan). In *Gabon*, Mobil, Elf, the Union Oil Group, Shell, Texaco,

* Based on Stanley H. Ruttenberg and Associates, *The American Oil Industry, A Failure of Anti-Trust Policy* (New York: Marine Engineers' Beneficial Association, 1973), "Appendix I, Selected Joint Ventures in the Oil Industry, by Regions of World, March 1973." North America not included.

Petrofina, BP, Exxon and others cloaked with the name Societe Equatoriale de Raffinage, operate a refinery that produces 850.000 tons of oil products a year.

In *Mozambique*, Sunray DX (Sun), Skelly and Clark are engaged in an oil exploration venture, plundering 15 million acres both offshore and onshore.

SOUTH AMERICA

In *Venezuela*, Sun, Atlantic Richfield and Texaco operate a gas compression plant that handles 150 million cubic feet of gas a day as well as an ammonia plant with a capacity of 1,500 tons a day. Here, Sun, Texaco, Phillips and Shell run a natural gas liquids plant. In addition, Mobil and Texaco have formed the South American Gulf Oil Co. which makes crude oil transporters and Gulf and Texaco together operate under the name of the Venezuela Gulf Refining Co.

In *Colombia*, Cities Service, Atlantic Richfield, Standard Oil of Indiana and Ecopetrol have formed the Colombia-Cities Service Petroleum Company here. In *Ecuador*, these two companies jointly engage in exploration and drilling.

EUROPE

In the *United Kingdom*, Standard Oil of California and British Petroleum Shell and BP, through United Kingdom Oil Pipeline, have teamed up to form BP-California Ltd. to run an oil pipeline. In *England*, Shell and BP have formed the Consolidated Petroleum Co., Ltd., a holding company. Also, Marathon, Continental Oil and Envoy Oil Ltd. have joined together for the exploration and production of oil and natural gas.

In *Ireland*, Exxon, Shell and Texaco, under the name of the Irish Refining Co., operate a refinery and Shell and BP have formed Irish Shell and BP. In *Scotland*, Shell and BP have formed the Shell and BP Scotland, Ltd.

In *Cyprus*, Mobil, BP and Shell operate a refinery under the name of the Cyprus Petroleum Refinery Ltd. In *Greece*, Mobil, Shell, Hellenic-Shipyards and the National Bank of Greece run a refinery under the name of the Hellenic Petroleum Refining Co.

In *West Germany*, BP,Exxon Erdol-Raffinerie Duisburg, Union Rheinische Barun-Kohin-Kraftsoff and Veba-Chemie run a pipeline under the name of Nord-West Oelleitung

Gmbtt. Also, Veba-Chemie, Texaco and Continental Oil run a refinery under the name of Oberrheinishe Mineraloelwerke Gmbtt.

In the *Netherlands*, Shell and Exxon, through N.V. Nederlandse Aardolie Mij. (NAM) engage in drilling, exploring and selling of crude oil and natural gas and Shell, Mobil, Standard Oil of California, Texaco and a Dutch company operate a pipeline between Rotterdam and the Rhine Basin in Germany under the name of N.V. Rotterdam Rijn Pijpleiding Mij. Also Royal Dutch Shell, Exxon, and BP jointly operate a one billion guilder tank complex for the production of crude oil and oil products.

In *Switzerland*, Shell and Gulf own a refinery under the name of Raffinerie de Cressier, S.A. In *Austria*, Rafinerie du Sud-Ouest, S.A., a consortium including BP, AGIP and Texaco produce crude oil and natural gas.

In *Italy*, Sarpom, consisting of Exxon, Texaco and Standard Oil of California run a refinery.

In *Luxembourg*, Texaco and Standard Oil of California operate through Texaco Luxembourg, S.A. In *Austria*, Exxon, Shell, BP, Mobil, Texaco, Marathon, Continental, EENI, Gelsenberg, Veba-Chemie, Wintershall and CFP operate 289 miles of 40 inch pipeline serving Austria and West Germany through the name of Transalpine Pipeline (TAL). In the *North Sea*, Standard Oil of Indiana, the British Gas Council, Amerada Hess and the Eastern Transmission Corp. are engaged in a 70 million dollar natural gas development program.

ASIA AND AUSTRALIA

In the *Philippines*, Exxon and Mobil through the Bataan Refining Corp. operate a refinery and Texaco, Standard Oil of California, Royal Dutch Shell, Exxon and Getty run a lubricating oil refinery.

In *Okinawa*, Standard Oil of California and Texaco operate a refinery. In *Japan*, Mobil, Exxon and a Japanese firm run a refinery under the name of Toa Nenryo Kogyo Kabushiki Kaisha.

In *Indonesia*, Exxon and Mobil, through its P.T. Stanvac Indonesia, engage in oil exploration and production.

In *New Zealand*, BP and Shell operate through BP-Shell Aquitaine & Todd Petroleum Development; also through

Shell-BP Pipeline Services, and Shell, BP and Todd Oil Services Ltd.

In *Australia*, Oil Search ltd., BP Group and Mobil Oil are engaged in oil exploration under the name of Australasian Petroleum Co. Proprietary Ltd. Mobil, BP Group and Interstate Oil Ltd. explore for natural gas through Frome-Broken Hill Co. pty., Ltd. The Island Exploration Co. Pty. Ltd., consisting of BP, Mobil, Oil Search Ltd. explore for oil and a consortium of Standard Oil of California, Texaco, Shell, Ampol Exp., Ltd. called the West Australian Petroleum Pty., Ltd. hold oil exploration rights covering 213,000 square miles.

MIDDLE EAST

In *Turkey*, Anadolu Tasfiyehanesi, A.S., consisting of Mobil, Shell and BP operate a refinery. In *Iraq*, the Iraq Petroleum Co. (IPC), including BP, Shell, CFP, Exxon, and Mobil explore, produce, transport and refine oil and oil products.

In *Saudi Arabia*, Aramco Overseas Co., including Texaco, Exxon, Standard Oil of California and Mobil, and the Aramco Realty Co., involving Texaco, Exxon, Standard Oil of California are located. Also, a consortium of Exxon, Texaco, Standard Oil of California and Mobil operate a pipeline under the name of the Trans Arabian Pipeline Co.

In *Iran*, CEP, Atlantic Richfield, Cities Service, Superior, Kerr-McGee, Sun and the National Iranian Oil Co. run the Iranian Offshore Petroleum Co., and BP, Shell, Gulf, Mobil, Exxon, Texaco and Standard Oil of California, calling themselves the Iranium Oil Consortium, explore, produce, transport and refine oil and oil products. Also, the Lavaan Petroleum Co., consisting of Atlantic Richfield, Murphy Oil, Union Oil and the National Iranian Co. explore and produce oil.

In *Kuwait*, BP, Gulf and Petrochemical Industries Co. operate under the name of the Kuwait Chemical Fertilizer Co., and Gulf and BP, under the name of the Kuwait Oil Co., Ltd. Here a consortium consisting of Gulf, BP and the Kuwait government operate a 30 million dollar gas facility.

In *Quatar*, a group consisting of BP, Shell, CFP, Mobil, and Exxon call themselves the Quatar Petroleum Co., Ltd. In the *Arabian Gulf*, Continental Oil, BP, Texaco and Sun are engaged in the development of the Fateh oil fields.

APPENDIX • C

A "first inning box-score" of trade union organizations and trade union leaders who, by March 1974, had already spoken up for some form of government take-over of the oil and other energy industries.* This appendix also includes the text of a bill for the public ownership of energy industries introduced in the State Legislature of Minnesota.

1. National Unions or Union Leaders, Affiliates, Districts, or Locals in Favor of Nationalization.

Amalgamated Meat Cutters and Butcher Workmen, AFL-CIO

Patrick E. Gorman, the union's chief executive officer, has proposed that the U.S. Congress consider methods of taking over and operating the oil industry (The Butcher Workmen, February 1974, p. 5.).

Service Employees International Union, AFL-CIO

George Hardy, president of the union, has issued a call to its locals urging that the U.S. oil industry be nationalized (Daily World, 13 February 1974.).

American Federation of State, County and Municpal Employees, AFL-CIO

Jerry Wurf, president of the union, has urged public ownership and operation of the oil industry ("Rank & File Currents" monthly newsletter of the New York Committee for Trade Union Action and Democracy, February 1974.).

* Compiled by the Labor Research Association, New York, 1974.

The National Hospital Union, a Division of the Retail, Wholesale and Department Store Union, AFL-CIO

In November, 1973, the founding convention of the National Hospital Union (of which Local 1199 is a major affiliate) adopted a resolution calling for the nationalization of the energy industry (Statement, Public Relations Department, Local 1199, 11 March 1974.).

Canada United Electrical Workers (Independent)

C.S. Jackson, president of the Canadian union, has sent a telegram to Canada's prime minister call for the nationalization of oil and energy production and distribution (UE News, 17 December, 1974.).

UE District 11, headed by Ernest DeMao, adopted a resolution calling for nationalization of the oil industry at a meeting in Milwaukee on February 9-10, 1974 (UE News, 25 February 1974.).

National Coordinating Committee for Trade Union Action and Democracy (Rank and File, Independent)

Fred Gaboury, field organizer, has stated that "the energy industry should be taken over and run as a public utility in the interest of the people" (Statement, 8 March 1974.).

2. National Unions or Union Leaders, Affiliates, Districts, or Locals in Favor of a TVA-type Organization to Compete with the Oil Barons

International Union of Electrical Radio and Machine Workers, AFL-CIO

Bill Bywater, president of District 3 (the union's largest), has said "We demand establishment of a TVA-type control by Congress of oil and other energy sources. If that doesn't work, we ought to nationalize the oil industry and utilities". (IUE News, February 1974.).

Paul Jennings, president of the union, in a message to members of the union's executive board, presidents of local unions and staff that was sent in January, demanded the creation of TVA-type federal agencies to: "seek and develop oil and natural gas sources (including oil shale) on public lands . . ." (Ibid.).

Automobile, Aerospace and Agricultural Implement Workers of America; International Union, United (UAW, Independent)

The union advocates a "public corporation answerable to the consumers and acting as a yardstick to keep the private companies honest . . ." (See UAW Washington Report, March 4, 1974.). Its 37-page National Energy Program calls for the expansion of "public agencies in competition with the private sector . . ." (p. 15).

Amalgamated Clothing Workers, AFL-CIO

The Union's general executive board at a meeting in San Diego, California, in February, advocated the "formation of a government corporation to explore, develop, and produce oil and gas from our huge reserves on public lands. (Advance, February 1974).

American Federation of Labor and Congress of Industrial Organizations

The AFL-CIO Executive Council met in Bel Harbour, Florida, on February 22 and adopted a resolution calling for the "creation of TVA-type fuel development agencies." A proposal for a take-over of the oil industry was not included in the resolution. However, "George Meany, president of the AFL-CIO proposed consideration of a special meeting to draw up a nationalization program and his suggestion was unanimously accepted by the council members, according to witnesses at the meeting". (New York Times, 23 February 1974.).

3. An important bill for public ownership of energy industries by the State of Minnesota has been introduced into the State's legislature, by Representatives Ojala, Fugina, D. Johnson, Jaros, and Rice—men from the Mesabi Range iron mining country, and from the central city working class areas of the Twin Cities. Their bill reflects the generations of anti-monopoly struggle of the Minnesota iron miners, trade unionists, and organized farmers. It is based on the correct finding that private interests "can no longer provide all forms of energy" at a reasonable cost and adequate amounts. The state is authorized to acquire any or all energy establishments either by negotiated purchase or by exercise of the right of eminent domain, which could be used to overcome the resistance of a company not wishing to sell.

While the bill has certain weaknesses, leaving it up to the governor to decide what to take over, and implying payment of high prices for the energy properties, it is a step forward.

The Text of the bill, H.F. No. 3390, follows in full:

A bill for an act

relating to the state; allowing the state to purchase or condemn and operate private industries and facilities that produce or distribute energy; providing for the issuance of bonds; appropriating money.

BE IT ENACTED BY THE LEGISLATURE OF THE STATE OF MINNESOTA:

Sec. 1. [PURPOSE.] The legislature finds that private industry and interests can no longer provide all forms of energy, including electricity, gas, oil and natural gas, at a reasonable cost to the consumer and in an amount adequate to the needs of the state and can no longer control the distribution of energy so that allocations are equitable. It is, therefore, the purpose of this act to allow for the state's acquisition and operation of energy facilities.

Sec. 2. [ACQUISITION; OPERATION OF FACILITIES.] The governor may acquire the property, real or personal, of any privately owned industry or facility that produces, manufactures or distributes energy, in any form, including electricity, gas, oil and natural gas, by purchase or eminent domain proceedings in the manner provided by law. Upon acquisition the governor may operate the facility on behalf of the state in a manner to be determined by the governor to provide and sell energy. Energy may be sold and distributed in a manner and at rates to be determined by the governor. The state may issue general obligation or revenue bonds in the manner provided by law in the amount of the appropriation in section 3 of this act.

Sec. 3. [APPROPRIATION.] The sum of $.......... is appropriated from the general fund to the governor for the purpose of acquiring and operating facilities pursuant to this act.

ACKNOWLEDGEMENTS

Any book worth reading has to be a product of collective thought. In writing this book I have greatly benefited from the numerous contributions of my co-workers, members of the National Council of the Communist Party, including their relatives and friends.

The idea of a book on the problems arising from the energy crisis came from my closest co-worker, Henry Winston, the National Chairman of our Party. Arnold Johnson and Lou Diskin made very useful editorial suggestions. I have been very fortunate to have been able to pluck the fruits of the economic and scientific researches of Victor Perlo, Warren E. Gillespie and Robert Phillipoff, the latter having been helpful for his overall commentary. Carol Marks has been of outstanding assistance. Her editorial and typing talents have been invaluable.

Most of all I am indebted to the numerous suggestions, criticisms, clippings and other information that I have received from many individuals across the land. Their names cannot all be mentioned but this book has made public their ideas.

INDEX

ABOUT THE AUTHOR

Gus Hall has been General Secretary of the Communist Party, U.S.A. since 1959. Born in Iron, Minnesota in 1910, he has spent his life in the working-class and Communist movements. He has worked as logger, miner, construction hand, mechanic, and steel worker. A founder of the United Steelworkers of America, he led the Little Steel strike (Ohio) in 1937. He served in the United States Navy in World War II.

Together with eleven other leaders of the Communist Party, Gus Hall was convicted in 1949 in a frame-up trial under the infamous, unconstitutional Smith Act. He served an eight-year term at Leavenworth—a sentence extended by a vindictive contempt of court citation.

Communist Party candidate for president in 1972, Gus Hall is recognized internationally as an outstanding Marxist-Leninist leader, theoretician and activist. He is the author of *Imperialism Today* (International Publishers, 1972).

